WebRTC Cookbook

Get to grips with advanced real-time communication
applications and services on WebRTC with practical,
hands-on recipes

Andrii Sergiienko

[PACKT] open source✲
PUBLISHING   community experience distilled

BIRMINGHAM - MUMBAI

WebRTC Cookbook

First published: February 2015

Production reference: 1200215

Published by Packt Publishing Ltd.
Livery Place
35 Livery Street
Birmingham B3 2PB, UK.

ISBN 978-1-78328-445-0

www.packtpub.com

Credits

Author
Andrii Sergiienko

Reviewers
Pasquale Boemio

Jose López

Marcos de Vera Piquero

Commissioning Editor
Usha Iyer

Acquisition Editor
Sam Wood

Content Development Editor
Rahul Nair

Technical Editor
Siddhi Rane

Copy Editor
Neha Vyas

Project Coordinator
Judie Jose

Proofreaders
Ting Baker

Simran Bhogal

Maria Gould

Paul Hindle

Indexer
Hemangini Bari

Production Coordinator
Melwyn Dsa

Cover Work
Melwyn Dsa

About the Author

Andrii Sergiienko is a computer software developer from Ukraine and is passionate about information technologies. From his early childhood, he was interested in computer programming and hardware. He took his first step into these fields more than 20 years ago. He has experience of a wide set of languages and technologies including C, C++, Java, assembly language, Erlang, JavaScript, PHP, Riak, shell scripting, computer networks, and security.

During his career he worked for both small, local companies such as domestic ISP and large, worldwide corporations such as Hewlett Packard. He also started his own projects—some of them were relatively successful.

Today, he is the owner and inspirer of OSLIKAS OÜ, a computer software company with headquarters in Estonia. The company (http://www.oslikas.com) focuses on modern IT technologies and solutions.

Working on this book was a really great and interesting experience for me. All this would be impossible without the help of certain people. And now is the time for me to say thank you to them.

First of all, I would like to thank my parents Olga and Alexander for my happy childhood that established the foundation for my life and career.

I would like to say thank you to my wife Inna for her patience, encouragement, and support during this process.

I would like to thank the Packt Publishing team as well. These guys are doing really great work and making the world a better place. We contacted some of them directly during the work, and others stayed behind the scenes. However, I know that a lot of people spent part of their lives to make this book possible. Thank you all.

About the Reviewers

Pasquale Boemio fell in love with Linux and the open source philosophy at the age of 12. Following this passion, he studied computer engineering at University of Naples Federico II from where he graduated with a master's degree.

Currently, he is working as a researcher in the Department of Electrical Engineering and Information Technology (DIETI) in the University of Naples Federico II, contributing to the development of real-time communication technologies. His efforts in this field are concretized by supporting the Meetecho project (www.meetecho.com).

Meetecho is a university spin-off and a tool for the collaborative work currently used by the Internet Engineering Task Force (IETF) to provide remote participation to the working groups. Meetecho leverages some state-of-the-art technologies (such as WebRTC and Docker) to implement a comprehensive architecture that can be lightweight and portable. Meetecho's best project is the Janus WebRTC Gateway (http://janus.conf.meetecho.com/), mentioned later in this book, which allows a user the ability to integrate different, real-time technologies without any pains.

In his spare time, Pasquale works on some personal open source projects (https://github.com/helloIAmPau) and helps the community by giving his contributions to cool projects found on the GitHub platform.

He has already worked with Packt Publishing by reviewing *WebRTC Integrator's Guide*, a useful guide for anyone who needs to integrate WebRTC with a retro technology such as SIP.

Jose López was born in Galicia, Spain. He is a telecommunications engineer with a large amount of experience in software development, and is also focused on real-time audio/video communications. He started working for Quobis Networks in 2013, a leading company in WebRTC solutions.

Marcos de Vera Piquero is a software engineer who has mainly worked with Python and CoffeeScript. His area of development is now focused on the server side of real-time multimedia applications at Quobis, his current employer. He's also a free software enthusiast and is trying to make it a real alternative.

www.PacktPub.com

Support files, eBooks, discount offers, and more

For support files and downloads related to your book, please visit www.PacktPub.com.

Did you know that Packt offers eBook versions of every book published, with PDF and ePub files available? You can upgrade to the eBook version at www.PacktPub.com and as a print book customer, you are entitled to a discount on the eBook copy. Get in touch with us at service@packtpub.com for more details.

At www.PacktPub.com, you can also read a collection of free technical articles, sign up for a range of free newsletters and receive exclusive discounts and offers on Packt books and eBooks.

https://www2.packtpub.com/books/subscription/packtlib

Do you need instant solutions to your IT questions? PacktLib is Packt's online digital book library. Here, you can search, access, and read Packt's entire library of books.

Why Subscribe?

- Fully searchable across every book published by Packt
- Copy and paste, print, and bookmark content
- On demand and accessible via a web browser

Free Access for Packt account holders

If you have an account with Packt at www.PacktPub.com, you can use this to access PacktLib today and view 9 entirely free books. Simply use your login credentials for immediate access.

Table of Contents

Preface

WebRTC is a relatively new and revolutionary technology that opens new horizons in the area of interactive applications and services. Most of the popular web browsers support it natively (such as Chrome and Firefox) or via extensions (such as Safari). Mobile platforms such as Android and iOS allow you to develop native WebRTC applications.

This book covers a wide set of topics on how to develop software using a WebRTC stack. Using practical recipes, it considers basic concepts, security, debugging, integration with other technologies, and other important themes of the development process in a friendly manner.

You will not only learn about WebRTC-specific features, but also attendant technologies (CSS3, HTML5, and WebSockets), and how to use them along with WebRTC.

What this book covers

Chapter 1, Peer Connections, introduces you to the very basic concepts of WebRTC. This includes practical recipes on peer connections. You will also find simple demo applications in this chapter.

Chapter 2, Supporting Security, leads you through various security-related topics and covers how to secure a typical WebRTC application's infrastructure components: SSL/TLS certificates, WebSockets, web servers, STUN/TURN, data channels, and more.

Chapter 3, Integrating WebRTC, considers integrating a WebRTC application with other technologies and third-party software. This chapter describes practical cases and solutions on integration.

Chapter 4, Debugging a WebRTC Application, is dedicated to application debugging—an important topic of the software development process. In this chapter, you will learn about the topics relating to debugging in the scope of WebRTC.

Chapter 5, Working with Filters, teaches you how to use CSS3 filters with WebRTC applications. This chapter also covers custom image processing.

Chapter 6, Native Applications, contains practical, step-by-step recipes dedicated to developing native WebRTC applications on mobile platforms.

Chapter 7, Third-party Libraries, describes general use cases and practical solutions based on third-party WebRTC frameworks and services.

Chapter 8, Advanced Functions, covers how to use advanced WebRTC features. It contains practical recipes on file transferring, streaming, audio/video controlling, and more.

What you need for this book

To use the recipes and codes provided and considered in this book, you will need a few pieces of software installed:

- ▶ Java SE 7: Note that for Android-related recipes from *Chapter 6, Native Applications*, you need Java SE 6 as well—the installation and configuration process is described in detail in this chapter.
- ▶ Erlang OTP 17: If you're familiar with this programming language, you can use this. If not, you can skip it—all Erlang examples are also provided in Java.
- ▶ Mac OS X and Xcode: Use this for recipes dedicated to developing WebRTC applications on iOS.
- ▶ Android and iOS: Use this for *Chapter 6, Native Applications*, which covers how to develop WebRTC applications for mobile platforms.
- ▶ Linux: Ubuntu is recommended.
- ▶ Chrome and Firefox: These are still the most WebRTC-friendly web browsers.

Specific requirements and configurations along with suggested solutions are considered in particular chapters.

Who this book is for

This book is written as a set of ready-to-use, practical recipes that cover a variety of topics related to developing WebRTC applications and services. It is assumed that you are familiar, in general, with WebRTC and its basic concepts.

Most of the provided recipes are written in JavaScript. However, server-side parts of applications are implemented in Erlang and Java. So, you are assumed to have at least basic experience with one of these technologies.

Working on some cases described in this book, you will have to deal with a Linux-based OS. All recipes are provided as a step-by-step guide. Although, if you have experience of working with and configuring Linux-based boxes, it would be useful.

So, this book is for someone who is familiar, in general, with the WebRTC stack, and who has at least basic skills in software development.

Conventions

In this book, you will find a number of styles of text that distinguish between different kinds of information. Here are some examples of these styles, and an explanation of their meaning.

Code words in text, database table names, folder names, filenames, file extensions, pathnames, dummy URLs, user input, and Twitter handles are shown as follows: "We can include our custom JavaScript library located in the `mylib.js` file."

A block of code is set as follows:

```
-module(sigserver_app).
-behaviour(application).
-export([start/2, stop/1, start/0]).
start() ->
    ok = application:start(ranch),
    ok = application:start(crypto),
    ok = application:start(cowlib),
    ok = application:start(cowboy),
```

When we wish to draw your attention to a particular part of a code block, the relevant lines or items are set in bold:

```
private static Map<Integer,Set<WebSocket>> Rooms = new HashMap<>();
    private int myroom;
    public Main() {
        super(new InetSocketAddress(30001));
    }
```

Any command-line input or output is written as follows:

```
rebar create-app appid=sigserver
```

New terms and **important words** are shown in bold. Words that you see on the screen, in menus or dialog boxes for example, appear in the text like this: "When the customer enters a message and clicks on the **Submit query** button, we will wrap the message into a JSON object and send it via the data channel."

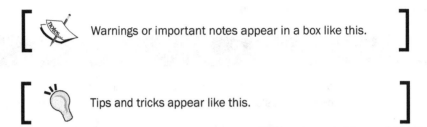

Warnings or important notes appear in a box like this.

Tips and tricks appear like this.

Reader feedback

Feedback from our readers is always welcome. Let us know what you think about this book—what you liked or disliked. Reader feedback is important for us as it helps us develop titles that you will really get the most out of.

To send us general feedback, simply e-mail feedback@packtpub.com, and mention the book's title in the subject of your message.

If there is a topic that you have expertise in and you are interested in either writing or contributing to a book, see our author guide at www.packtpub.com/authors.

Customer support

Now that you are the proud owner of a Packt book, we have a number of things to help you to get the most from your purchase.

Downloading the example code

You can download the example code files from your account at http://www.packtpub.com for all the Packt Publishing books you have purchased. If you purchased this book elsewhere, you can visit http://www.packtpub.com/support and register to have the files e-mailed directly to you.

Downloading the color images of this book

We also provide you with a PDF file that has color images of the screenshots/diagrams used in this book. The color images will help you better understand the changes in the output. You can download this file from: `https://www.packtpub.com/sites/default/files/downloads/4450OS_ColoredImages.pdf`.

Errata

Although we have taken every care to ensure the accuracy of our content, mistakes do happen. If you find a mistake in one of our books—maybe a mistake in the text or the code—we would be grateful if you could report this to us. By doing so, you can save other readers from frustration and help us improve subsequent versions of this book. If you find any errata, please report them by visiting `http://www.packtpub.com/submit-errata`, selecting your book, clicking on the **Errata Submission Form** link, and entering the details of your errata. Once your errata are verified, your submission will be accepted and the errata will be uploaded to our website or added to any list of existing errata under the Errata section of that title.

To view the previously submitted errata, go to `https://www.packtpub.com/books/content/support` and enter the name of the book in the search field. The required information will appear under the **Errata** section.

Piracy

Piracy of copyrighted material on the Internet is an ongoing problem across all media. At Packt, we take the protection of our copyright and licenses very seriously. If you come across any illegal copies of our works in any form on the Internet, please provide us with the location address or website name immediately so that we can pursue a remedy.

Please contact us at `copyright@packtpub.com` with a link to the suspected pirated material.

We appreciate your help in protecting our authors and our ability to bring you valuable content.

Questions

If you have a problem with any aspect of this book, you can contact us at `questions@packtpub.com`, and we will do our best to address the problem.

1

Peer Connections

In this chapter, we will cover the following topics:

- ▸ Building a signaling server in Erlang
- ▸ Building a signaling server in Java
- ▸ Detecting WebRTC functions supported by a browser
- ▸ Making and answering calls
- ▸ Implementing a chat using data channels
- ▸ Implementing a chat using a signaling server
- ▸ Configuring and using STUN
- ▸ Configuring and using TURN

Introduction

This chapter covers the basic concepts of how to use WebRTC when developing rich media web applications and services.

With simple and short recipes, you will learn how to create your own signaling server. The key data that needs to be exchanged by peers before they establish a direct connection is called the **session description**—it specifies the peers' configuration. Signaling server is a component in an application's infrastructure that is accessible by all peers and serves to exchange multimedia's session description. The way peers should exchange data is not described by WebRTC standards, so you should make the decision on your own regarding the protocol and mechanism you will use for this task.

You can build a signaling server using any programming language and technology you like. In general, the signaling protocol can be non-technical and is possible to implement in away where the peers would use just a sheet of paper to exchange necessary data between each other. In this chapter, we use WebSocket to implement signaling, although you can use any other protocol.

The signaling stage is represented in the schema that is shown in the following diagram:

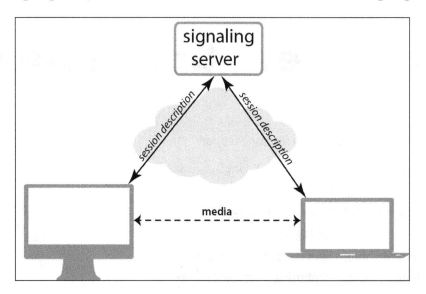

In this chapter, you will find two recipes that are dedicated to signaling server development: *Building a signaling server in Erlang* and *Building a signaling server in Java*. Java is probably the most popular and known technology, and it would be easy to get into this topic using Java, even if you don't have programming experience with this technology. Erlang is not widely known yet. Nonetheless, this is a very mature technology, very suitable for writing lightweight and extremely fast server applications with perfect scalability. So, by learning signaling server, you will find simple solutions in Erlang as well.

This chapter also covers the basic use case of how to use WebRTC data channels: file transferring and peer-to-peer chat.

You will also learn how to configure and use **Session Traversal Utilities for NAT** (**STUN**) and **Traversal Using Relays around NAT** (**TURN**) services, and of course, this chapter covers making peer-to-peer calls using WebRTC.

Note that in this chapter, we will cover the process of making computer-to-computer calls. If you want to know more about how to use WebRTC with VoIP and SIP, and how to make phone calls from a web page, refer to the *Chapter 3, Integrating WebRTC*.

Building a signaling server in Erlang

The following recipe shows how to build signaling server using Erlang programming language and WebSockets for transport protocol. We will not introduce Erlang programming in this recipe, so you should have at least basic knowledge of this programming language and its relevant technologies.

Getting ready

To use this solution, you should have Erlang installed on your system to start with. You can download the Erlang distribution relevant to your system from its home page http://www.erlang.org/download.html. The installation process might need specific actions relevant to specific platforms/OSes, so follow the official installation instructions at http://www.erlang.org/doc/installation_guide/INSTALL.html.

 For this example, I've used Erlang 17. You might need to add some minor changes to the code to make it work under future versions of Erlang.

We will also use the Git versioning system to download additional packets and components necessary for our solution, so you should download and install Git distribution relevant to your system. You can download this from http://www.git-scm.com. As a build tool for the project, we will use Rebar; you should also download and install it from https://github.com/basho/rebar.

How to do it...

The following steps will lead you through the process of building a signaling server using Erlang:

1. Create a new folder for the signaling server application and navigate to it.

2. Using the Rebar tool, create a basic Erlang application:

    rebar create-app appid=sigserver

    This command will create an `src` folder and relevant application files in it.

3. Create the `rebar.config` file, and put the following Rebar configuration in it:

    ```
    {erl_opts, [warnings_as_errors]}.
    {deps,
    [
    {'gproc', ".*", {
    git, "git://github.com/esl/gproc.git", {tag, "0.2.16"}
    }},
    ```

```
{'jsonerl', ".*", {
git, "git://github.com/fycth/jsonerl.git", "master"
}},
{'cowboy', ".*", {
git,"https://github.com/extend/cowboy.git","0.9.0"
}}
]}.
```

4. Open the `src/sigserver.app.src` file and add the following components to the application's section list: `cowlib`, `cowboy`, `compiler`, and `gproc`.

5. Open the `src/sigserver_app.erl` file and add the following code:

```
-module(sigserver_app).
-behaviour(application).
-export([start/2, stop/1, start/0]).
start() ->
    ok = application:start(ranch),
    ok = application:start(crypto),
    ok = application:start(cowlib),
    ok = application:start(cowboy),
    ok = application:start(gproc),
    ok = application:start(compiler),
    ok = application:start(sigserver).

start(_StartType, _StartArgs) ->
    Dispatch = cowboy_router:compile([
              {'_',[
                 {"/ ", handler_websocket,[]}
               ]}
    ]),
    {ok, _} = cowboy:start_http(websocket, 100, [{ip,
    {127,0,0,1}},{port, 30001}], [
            {env, [{dispatch, Dispatch}]},
            {max_keepalive, 50},
            {timeout, 500}]),
    sigserver_sup:start_link().

stop(_State) -> ok.
```

6. Create the `src/handler_websocket.erl` file and put the following code in it:

```
-module(handler_websocket).
-behaviour(cowboy_websocket_handler).
-export([init/3]).
-export([websocket_init/3, websocket_handle/3,
        websocket_info/3, websocket_terminate/3]).
```

```erlang
-record(state, {
        client = undefined :: undefined | binary(),
        state = undefined :: undefined | connected |
        running,
        room = undefined :: undefined | integer()
}).

init(_Any, _Req, _Opt) ->
    {upgrade, protocol, cowboy_websocket}.

websocket_init(_TransportName, Req, _Opt) ->
    {Client, Req1} = cowboy_req:header(<<"x-forwarded-
    for">>, Req),
    State = #state{client = Client, state = connected},
    {ok, Req1, State, hibernate}.

websocket_handle({text,Data}, Req, State) ->
    StateNew = case (State#state.state) of
                    started ->
                        State#state{state = running};
                    _ ->
                        State
               end,
    JSON = jsonerl:decode(Data),
    {M,Type} = element(1,JSON),
    case M of
        <<"type">> ->
            case Type of
                <<"GETROOM">> ->
                    Room = generate_room(),
                    R =
                    iolist_to_binary(jsonerl:encode({
                    {type, <<"GETROOM">>},
                    {value, Room}})),
                    gproc:reg({p,l, Room}),
                    S = (StateNew#state{room = Room}),
                    {reply, {text, <<R/binary>>}, Req, S,
                    hibernate};
                <<"ENTERROOM">> ->
                    {<<"value">>,Room} = element(2,JSON),
                    Participants =
                    gproc:lookup_pids({p,l,Room}),
                    case length(Participants) of
                        1 ->
```

```erlang
                                gproc:reg({p,l, Room}),
                                S = (StateNew#state{room =
                                Room}),
                                {ok, Req, S, hibernate};
                        _ ->
                                R =
                                iolist_to_binary(jsonerl:
                                encode({{type,
                                <<"WRONGROOM">>}})),
                                {reply, {text, <<R/binary>>},
                                Req, StateNew, hibernate}
                        end;
                _ ->
                        reply2peer(Data, StateNew#state.room),
                        {ok, Req, StateNew, hibernate}
                end;
        _ ->
                reply2peer(Data, State#state.room),
                {ok, Req, StateNew, hibernate}
        end;

websocket_handle(_Any, Req, State) -> {ok, Req, State, hibernate}.

websocket_info(_Info, Req, State) -> {reply, {text,_Info}, Req,
State, hibernate}.

websocket_terminate(_Reason, _Req, _State) -> ok.

reply2peer(R, Room) ->
    [P ! <<R/binary>> || P <- gproc:lookup_pids({p,l,Room})
    -- [self()]].

generate_room() ->
    random:seed(now()),
    random:uniform(999999).
```

7. Now we can compile the solution using the Rebar tool:

   rebar get-deps

   rebar compile

   If everything was successful, you should not see any errors (warnings are not critical).

8. After we build our signaling server, we can start it using the following command:

   erl -pa deps/*/ebin ebin -sasl errlog_type error -s sigserver_app

Windows-based systems can't use a star symbol in such constructions, so if you're working under Windows, you should use the full path name as shown in the following command:

```
erl -pa deps/cowboy/ebin deps/cowlib/ebin deps/gproc/
ebin deps/jsonerl/ebin deps/ranch/ebin ebin -sasl
errlog_type error -s sigserver_app
```

Now your signaling server should be running, and you need to listen for incoming WebSocket connections on port 30001.

Note that full source codes are supplied with this book.

Downloading the example code

You can download the example code files from your account at http://www.packtpub.com for all the Packt Publishing books you have purchased. If you purchased this book elsewhere, you can visit http://www.packtpub.com/support and register to have the files e-mailed directly to you.

How it works...

In this recipe, we implemented the WebRTC signaling server in Erlang. The application listens on port 30001 for incoming WebSocket connections from the browser clients.

The first peer will be registered by the server in a virtual room and will get the room number. The second peer after that can use the room number in order to connect to the first peer. The signaling server will check whether the virtual room exists and if so, it will route call/answer requests and answers between the peers in order to make them establish a direct peer-to-peer WebRTC connection.

There's more...

Basically, this is a very simple signaling server. It doesn't have any advanced features, and the main goal of it is to help peers establish direct connection between each other. Nevertheless, a signaling server can serve additional tasks. For example, it can serve for web chats, file transfers, service data exchanges, and other features specific for certain situations. There are no certain requirements for a signaling server; you can implement it using your favorite programming language and technology.

See also

▶ For tips on implementing a signaling server in Java, refer to the *Building a signaling server in Java* recipe

▶ You can also refer to the *Making and answering calls* recipe on how to use a signaling server from a browser application using JavaScript

Building a signaling server in Java

In this recipe, we will cover the implementation of a signaling server in Java.

Getting ready

This recipe uses Java, so you should have **Java Developer Kit** (**JDK**) installed on your machine. You can download the appropriate version of JDK for your platform from its web page at http://www.java.com.

Java 7 has its own API to implement a WebSocket application. Previous versions of Java don't have the native support of WebSockets. In this recipe, we will cover the universal solution that works in different Java versions and is based on the third-party component, which you can find on its home page at http://java-websocket.org. This project is also present on GitHub at https://github.com/TooTallNate/Java-WebSocket.

You need to download and install the Java-WebSocket library; it should then be linked to your project.

In this recipe, we pack signaling messages into the JSON format before sending, so we need a Java library to work with JSON structures. For this purpose, we will use Java classes from JSON's home page, http://www.json.org/java/.

Download these classes and link them to your project, or you can just put these classes into your project's folder structure and compile it all together.

It is assumed that you have experience of programming in Java, so we will not cover the basic questions like how to start a Java application and so on.

How to do it...

Create a new project in your Java IDE and link the JSON libraries along with the Java-WebSocket library.

The following code represents a simple signaling server. Compile it and start a Java console application as usual:

```java
package com.webrtcexample.signaler;

import org.java_websocket.WebSocket;
import org.java_websocket.handshake.ClientHandshake;
import org.java_websocket.server.WebSocketServer;
import org.json.JSONException;
import org.json.JSONObject;

import java.net.InetSocketAddress;
import java.util.*;

public class Main extends WebSocketServer {

    private static Map<Integer,Set<WebSocket>> Rooms = new
HashMap<>();
    private int myroom;

    public Main() {
        super(new InetSocketAddress(30001));
    }

    @Override
    public void onOpen(WebSocket conn, ClientHandshake handshake)
    {
        System.out.println("New client connected: " +
        conn.getRemoteSocketAddress() + " hash " +
        conn.getRemoteSocketAddress().hashCode());
    }

    @Override
    public void onMessage(WebSocket conn, String message) {

        Set<WebSocket> s;
        try {
            JSONObject obj = new JSONObject(message);
            String msgtype = obj.getString("type");
            switch (msgtype) {
                case "GETROOM":
                    myroom = generateRoomNumber();
                    s = new HashSet<>();
                    s.add(conn);
                    Rooms.put(myroom, s);
```

```
                              System.out.println("Generated new room: " +
                              myroom);
                              conn.send("{\"type\":\"GETROOM\",\"value\":" +
                              myroom + "}");
                              break;
                        case "ENTERROOM":
                              myroom = obj.getInt("value");
                              System.out.println("New client entered room "
                              + myroom);
                              s = Rooms.get(myroom);
                              s.add(conn);
                              Rooms.put(myroom, s);
                              break;
                        default:
                              sendToAll(conn, message);
                              break;
                  }
            } catch (JSONException e) {
                  sendToAll(conn, message);
            }
            System.out.println();
      }


      @Override
      public void onClose(WebSocket conn, int code, String reason,
      boolean remote) {
            System.out.println("Client disconnected: " + reason);
      }


      @Override
      public void onError(WebSocket conn, Exception exc) {
            System.out.println("Error happened: " + exc);
      }


      private int generateRoomNumber() {
            return new Random(System.currentTimeMillis()).nextInt();
      }


      private void sendToAll(WebSocket conn, String message) {
            Iterator it = Rooms.get(myroom).iterator();
            while (it.hasNext()) {
                  WebSocket c = (WebSocket)it.next();
                  if (c != conn) c.send(message);
            }
      }
```

```
public static void main(String[] args) {
    Main server = new Main();
    server.start();
}
}
```

Once the application starts, it will listen on the TCP port 30001 for WebSocket messages from clients. You can write simple client applications to test the signaling server—refer to the *Making and answering calls* recipe.

Note that you can find a Maven-based project for this example supplied with this book.

How it works...

First of all, the client sends a GETROOM message to the signaling server that is listening on TCP port 30001. The server generates a new virtual room number, stores it, and sends it back to the client.

The client constructs a new access URL using the virtual room number received from the server. Then, the second client uses this URL to enter the virtual room and establish a call to the first client.

The second client sends the room number it got from the URL to the signaling server. The server associates the client with the virtual room number. Then, the client makes a call, using signaling server, which forwards its messages to the first client that is present in the room already. The first client answers the call, also using the signaling server as the middle point.

So both clients exchange the necessary data (including network details) and then establish direct peer-to-peer connection. After the connection is established, peers don't use the server anymore.

There's more...

The WebSocket signaling server in Java can be implemented using a Java EE stack. For more details, take a look at the home page of JSR 356 at http://www.oracle.com/technetwork/articles/java/jsr356-1937161.html.

You can also find an example at https://github.com/hsilomedus/web-sockets-samples/tree/master/eesockets.

Another solution is to use Spring 4. It has WebSocket support out of the box. For details on this solution, take a look at the example on GitHub at https://github.com/hsilomedus/web-sockets-samples/tree/master/springsockets.

See also

▸ For an alternative solution, you can refer to the *Building a signaling server in Erlang* recipe

Detecting WebRTC functions supported by a browser

WebRTC is not fully supported by all available web browsers at this time. Moreover, there is a chance that your application will be running under some kind of exotic environment or web browser that does not support WebRTC. So you need to have some mechanism that would enable you to detect whether the environment in which your web application is running supports the necessary WebRTC features the application is going to use. In this recipe, we will cover the basic method of doing that.

Getting ready

This task is relevant for the client side only, so all the code will be written in JavaScript. Thus, no specific preparation is needed.

How to do it...

You can write a JavaScript library that can be used to detect which WebRTC methods are available under the environment and by what names they are known for your application.

The following code represents a basic but productive example of such a kind of library:

```javascript
var webrtcDetectedVersion = null;
var webrtcDetectedBrowser = null;
window.requestFileSystem = window.requestFileSystem || window.
webkitRequestFileSystem;

function initWebRTCAdapter() {
    if (navigator.mozGetUserMedia) {
        webrtcDetectedBrowser = "firefox";
        webrtcDetectedVersion =
        parseInt(navigator.userAgent.match(/Firefox\/
        ([0-9]+)\./)[1], 10);

        RTCPeerConnection = mozRTCPeerConnection;
        RTCSessionDescription = mozRTCSessionDescription;
        RTCIceCandidate = mozRTCIceCandidate;
        getUserMedia = navigator.mozGetUserMedia.bind(navigator);
```

```
attachMediaStream =
    function(element, stream) {
        element.mozSrcObject = stream;
        element.play();
    };

reattachMediaStream =
    function(to, from) {
        to.mozSrcObject = from.mozSrcObject;
        to.play();
    };

MediaStream.prototype.getVideoTracks =
    function() {
        return [];
    };

MediaStream.prototype.getAudioTracks =
    function() {
        return [];
    };
return true;
} else if (navigator.webkitGetUserMedia) {
webrtcDetectedBrowser = "chrome";
webrtcDetectedVersion =
parseInt(navigator.userAgent.match(/Chrom(e|ium)\/
  ([0-9]+)\./)[2], 10);

RTCPeerConnection = webkitRTCPeerConnection;
getUserMedia =
navigator.webkitGetUserMedia.bind(navigator);
attachMediaStream =
    function(element, stream) {
        element.src = webkitURL.createObjectURL(stream);
    };

reattachMediaStream =
    function(to, from) {
        to.src = from.src;
    };

if (!webkitMediaStream.prototype.getVideoTracks) {
    webkitMediaStream.prototype.getVideoTracks =
        function() {
```

```
                    return this.videoTracks;
            };
        webkitMediaStream.prototype.getAudioTracks =
            function() {
                return this.audioTracks;
            };
    }

    if (!webkitRTCPeerConnection.prototype.getLocalStreams) {
        webkitRTCPeerConnection.prototype.getLocalStreams =
            function() {
                return this.localStreams;
            };
        webkitRTCPeerConnection.prototype.getRemoteStreams =
            function() {
                return this.remoteStreams;
            };
    }
    return true;
} else return false;
};
```

How it works...

This solution tests which WebRTC API methods are available in the environment and how they are named. So your application can use certain API function names that will be relevant for any web browser, without using browser-specific function names.

There's more...

There is another way to solve this task. You don't necessary have to write your own adapter. You can take the adapter prepared by Google. It can be found at `http://apprtc.webrtc. org/js/adapter.js`. You just need to include it in your JavaScript code.

You can also consider using a browser's plugin that enables the use of WebRTC in Safari and Internet Explorer. You can get these at `https://temasys.atlassian.net/wiki/ display/TWPP/How+to+integrate+the+plugin+with+your+website`.

See also

You can find more information on the adapter at the web page `http://www.webrtc.org/ web-apis/interop`.

Making and answering calls

The very basic action of any WebRTC application is making and receiving a call. This recipe shows how to make calls to a remote peer.

Getting ready

At the beginning, peers don't know each other, and they don't know the necessary network information to make direct connection possible. Before establishing a direct connection, peers should exchange necessary data using some middle point—usually, a signaling server. This is a middle point that is known to each peer. So each peer can connect to the signaling server, and then one peer can call another one—by asking the signaling server to exchange specific data with another peer and make peers know each other.

So, you need a signaling server to run.

How to do it...

Before two peers can establish a direct connection, they should exchange specific data (ICE candidates and session descriptions) using a middle point—the signaling server. After that, one peer can call another one, and the direct peer-to-peer connection can be established.

Interactive Connectivity Establishment (**ICE**) is a technique used in **Network Address Translator** (**NAT**), which bypasses the process of establishing peer-to-peer direct communication. Usually, ICE candidates provide information about the IP address and port of the peer. Typically, an ICE candidate message might look like the following:

```
a=candidate:1 1 UDP 4257021352 192.168.0.10 1211 typ host
```

Session Description Protocol (**SDP**) is used by peers in WebRTC to configure exchanging (network configuration, audio/video codecs available, and so on). Every peer sends details regarding its configuration to another peer and gets the same details from it back. The following print depicts a part of an SDP packet representing the audio configuration options of a peer:

```
m=audio 53275 RTP/SAVPF 121 918 100 1 2 102 90 131 16

c=IN IP4 16.0.0.1

a=rtcp:53275 IN IP4 16.0.0.1
```

In the schema represented in the following diagram, you can see the generic flow of a call establishing process:

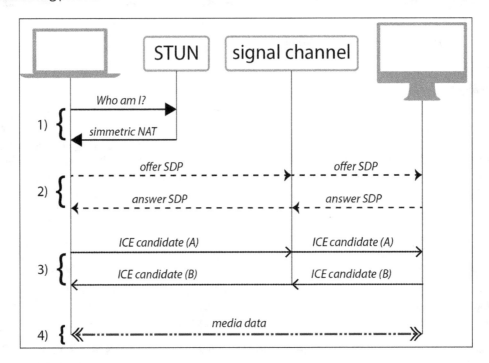

Note that TURN is not showed in the schema. If you used TURN, it would be depicted just after the STUN stage (before the first and second stage).

Making a call

To make a call, we need to take some steps to prepare (such as getting access to the browser's media):

1. Get access to the user's media:

```
function doGetUserMedia() {
    var constraints = {"audio": true, "video":
    {"mandatory": {}, "optional": []}};
        try {
            getUserMedia(constraints, onUserMediaSuccess,
                function(e) {
                    console.log("getUserMedia error "+
                    e.toString());
                });
        } catch (e) {
            console.log(e.toString());
        }
    };
```

2. If you succeed, create a peer connection object and make a call:

```
function onUserMediaSuccess(stream) {
        attachMediaStream(localVideo, stream);
        localStream = stream;
        createPeerConnection();
        pc.addStream(localStream);
        if (initiator) doCall();
};
function createPeerConnection() {
        var pc_constraints = {"optional":
        [{"DtlsSrtpKeyAgreement": true}]};
        try {
            pc = new RTCPeerConnection(pc_config,
            pc_constraints);
            pc.onicecandidate = onIceCandidate;
        } catch (e) {
            console.log(e.toString());
            pc = null;
            return;
        }
        pc.onaddstream = onRemoteStreamAdded;
};

function onIceCandidate(event) {
        if (event.candidate)
            sendMessage({type: 'candidate', label:
            event.candidate.sdpMLineIndex, id:
            event.candidate.sdpMid,
            candidate: event.candidate.candidate});
};

function onRemoteStreamAdded(event) {
        attachMediaStream(remoteVideo, event.stream);
        remoteStream = event.stream;
};

function doCall() {
        var constraints = {"optional": [], "mandatory":
        {"MozDontOfferDataChannel": true}};
        if (webrtcDetectedBrowser === "chrome")
            for (var prop in constraints.mandatory) if
            (prop.indexOf("Moz") != -1) delete
            constraints.mandatory[prop];
```

```
            constraints = mergeConstraints(constraints,
            sdpConstraints);
            pc.createOffer(setLocalAndSendMessage,
            errorCallBack, constraints);
    };
```

Answering a call

Assuming that we will use WebSockets as a transport protocol for exchanging data with signaling server, every client application should have a function to process messages coming from the server. In general, it looks as follows:

```
function processSignalingMessage(message) {
        var msg = JSON.parse(message);
        if (msg.type === 'offer') {
            pc.setRemoteDescription(new
            RTCSessionDescription(msg));
            doAnswer();
        } else if (msg.type === 'answer') {
            pc.setRemoteDescription(new
            RTCSessionDescription(msg));
        } else if (msg.type === 'candidate') {
            var candidate = new
            RTCIceCandidate({sdpMLineIndex:msg.label,
            candidate:msg.candidate});
            pc.addIceCandidate(candidate);
        } else if (msg.type === 'GETROOM') {
            room = msg.value;
            onRoomReceived(room);
        } else if (msg.type === 'WRONGROOM') {
            window.location.href = "/";
        }
    };
```

This function receives messages from the signaling server using the WebSockets layer and acts appropriately. For this recipe, we are interested in the offer type of message and doAnswer function.

The doAnswer function is presented in the following listing:

```
function doAnswer() {
    pc.createAnswer(setLocalAndSendMessage, errorCallBack,
    sdpConstraints);
};
```

The `sdpConstraints` object describes the WebRTC connection options to be used. In general, it looks as follows:

```
var sdpConstraints = {'mandatory': {'OfferToReceiveAudio':true,
'OfferToReceiveVideo':true }};
```

Here we can say that we would like to use both audio and video while establishing WebRTC peer-to-peer connection.

The `errorCallback` method is a callback function that is called in case of an error during the calling of the `createAnswer` function. In this callback function, you can print a message to the console that might help to debug the application.

The `setLocalAndSendMessage` function sets the local session description and sends it back to the signaling server. This data will be sent as an answer type of message, and then the signaling server will route this message to the caller:

```
function setLocalAndSendMessage(sessionDescription) {
    pc.setLocalDescription(sessionDescription);
    sendMessage(sessionDescription);
};
```

Note that you can find the full source code for this example supplied with this book.

How it works...

Firstly, we will ask the web browser to gain access to the user media (audio and video). The web browser will ask the user for these access rights. If we get the access, we can create a connection peer entity and send the call message to the signaling server, which will route this message to the remote peer.

The workflow of the code is very simple. The `processSignalingMessage` function should be called every time we get a message from the signaling server. Usually, you should set it as an `onmessage` event handler of the WebSocket JavaScript object.

After the message is received, this function detects the message type and acts appropriately. To answer an incoming call, it calls the `doAnswer` function that will do the rest of the magic—prepare the session description and send it back to the server.

The signaling server will get this reply as an answer message and will route it to the remote peer. After that, peers will have all the necessary data on each other to start establishing a direct connection.

There's more...

This is the basic functionality of WebRTC. Most of your applications will probably have the same code for this task. The only big difference might be communication with the signaling server—you can use any protocol you like.

See also

► Refer to the *Implementing a chat using data channels* recipe regarding the process of building a simple web chat application using WebRTC

► You can find more details on ICE on the RFC 5245 website at `https://tools.ietf.org/html/rfc5245`

► More information regarding SDP can be found on RFC 4566 at `https://tools.ietf.org/html/rfc4566`

Implementing a chat using data channels

In this recipe, we will implement a peer-to-peer private messaging service using WebRTC data channels. This method allows us to send messages directly from peer to peer, using secure and safe data channels provided by the WebRTC stack.

The schema represented in the following diagram depicts the generic feature flow:

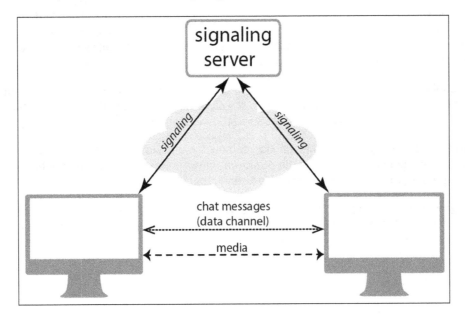

Getting ready

We will develop a simple application, so you don't need any specific preparations for this recipe. A signaling server is necessary for this application, and it can be taken from the *Building a signaling server in Erlang* or *Building a signaling server in Java* recipe.

How to do it...

For simplicity, we will make two parts of the application: an index web page and a JavaScript library.

Creating the main HTML page of the application

1. First, create an HTML `index.html` page. In the following code, you can find its content. Note that the less important and obvious parts might be skipped here.

   ```
   <!DOCTYPE html>
   <html>
   <head>
   ```

2. Include our JavaScript library that is in a separate file:

   ```
   <script type="text/javascript" src="myrtclib.js"></script>
   ```

3. Include Google's WebRTC JavaScript adapter:

   ```
   <script src="https://rawgit.com/GoogleChrome/webrtc/master/
   samples/web/js/adapter.js"></script>
   </head>
   <body>
   ```

4. Create a `div` tag where we will put information regarding the connection:

   ```
   <div id="status"></div><br>
   ```

5. Create a `div` tag where the received messages from a remote peer will be placed:

   ```
   <div id="chat"></div>
   ```

6. Create a form with an input element and a button to send messages to the remote peer:

   ```
   <form name="chat_form" onsubmit="onChatSubmit(document.chat_form.
   msg.value);
   return false;">
       <input type="text" class="search-query"
       placeholder="chat here" name="msg" id="chat_input">
       <input type="submit" class="btn" id="chat_submit_btn"/>
   </form>
   <script>
   ```

7. Create a connection to the signaling server and initialize the WebRTC stack. The following function is declared in the JavaScript library, which we will consider further in the recipe:

```
myrtclibinit("ws://localhost:30001");
```

Note that the domain name and port might be different in your case; they should be the same as declared in the source codes of the signaling sever. By default, the signaling server is listening on local host and on port 30001.

The following function sends a message to the remote peer using the `sendDataMessage` function—we will write it as part of the JavaScript library:

```
function onChatSubmit(txt) {
    var msg = JSON.stringify({"type" : "chatmessage", "txt" :
    txt});
    sendDataMessage(msg);
};
```

We will also declare a callback function for a catching event when a new virtual room is created:

```
function onRoomReceived(room) {
    var st = document.getElementById("status");
```

Create a link to share with the remote peer, put the link in the `div` status.

```
    st.innerHTML = "Now, if somebody wants to join you, should use
    this link: <a href=\""+window.location.href+"?
    room="+room+"\">"+window.location.href+"?room="+room+"</a>";
};
```

To show the messages received from the remote peer, we will declare an appropriate callback function. This function gets the message and puts it in the appropriate place on the HTML page:

```
function onPrivateMessageReceived(txt) {
    var t = document.getElementById('chat').innerHTML;
    t += "<br>" + txt;
    document.getElementById('chat').innerHTML = t;
}
</script>
</body>
</html>
```

Save the HTML file. This will be the main page of the applications.

Creating the JavaScript helper library

Now, create an empty `myrtclib.js` file and put the following content into it. Note that many parts of the following code might be used in the next chapters, so they should be well-known to you already. Such obvious parts of the code might be skipped in further.

```
var RTCPeerConnection = null;
var room = null;
var initiator;
var pc = null;
var signalingURL;
```

The following variable will be used for handling the data channel object:

```
var data_channel = null;
var channelReady;
var channel;
var pc_config = {"iceServers":
    [{url:'stun:23.21.150.121'},
     {url:'stun:stun.l.google.com:19302'}]};

function myrtclibinit(sURL) {
    signalingURL = sURL;
    openChannel();
};

function openChannel() {
    channelReady = false;
    channel = new WebSocket(signalingURL);
    channel.onopen = onChannelOpened;
    channel.onmessage = onChannelMessage;
    channel.onclose = onChannelClosed;
};

function onChannelOpened() {
    channelReady = true;
    createPeerConnection();

    if(location.search.substring(1,5) == "room") {
        room = location.search.substring(6);
        sendMessage({"type" : "ENTERROOM", "value" : room *
        1});
        initiator = true;
        doCall();
    } else {
        sendMessage({"type" : "GETROOM", "value" : ""});
```

```
                    initiator = false;
            }
    };

    function onChannelMessage(message) {
        processSignalingMessage(message.data);
    };

    function onChannelClosed() {
        channelReady = false;
    };

    function sendMessage(message) {
        var msgString = JSON.stringify(message);
        channel.send(msgString);
    };

    function processSignalingMessage(message) {
        var msg = JSON.parse(message);

        if (msg.type === 'offer') {
            pc.setRemoteDescription(new
            RTCSessionDescription(msg));
            doAnswer();
        } else if (msg.type === 'answer') {
            pc.setRemoteDescription(new
            RTCSessionDescription(msg));
        } else if (msg.type === 'candidate') {
            var candidate = new
            RTCIceCandidate({sdpMLineIndex:msg.label,
            candidate:msg.candidate});
            pc.addIceCandidate(candidate);
        } else if (msg.type === 'GETROOM') {
            room = msg.value;
            onRoomReceived(room);
        } else if (msg.type === 'WRONGROOM') {
            window.location.href = "/";
        }
    };

    function createPeerConnection() {
        try {
            pc = new RTCPeerConnection(pc_config, null);
            pc.onicecandidate = onIceCandidate;
```

Until now, the code is very similar to what we used in a typical WebRTC example application. Although, now we will add something new. We will set up a handler for the `ondatachannel` event of the `PeerConnection` object. This callback function will be called when the peer asks us to create a data channel and establish a data connection:

```
        pc.ondatachannel = onDataChannel;
    } catch (e) {
        console.log(e);
        pc = null;
        return;
    }
};
```

The handler function is pretty simple. We will store the reference in the data channel and initialize it:

```
function onDataChannel(evt) {
    console.log('Received data channel creating request');
    data_channel = evt.channel;
    initDataChannel();
}
```

By initializing the data channel, I mean setting up a channel's event handlers:

```
function initDataChannel() {
    data_channel.onopen = onChannelStateChange;
    data_channel.onclose = onChannelStateChange;
    data_channel.onmessage = onReceiveMessageCallback;
}
```

In the following function, we need to create a new data channel—not when the remote peer is asking us, but when we're the initiator of the peer connection and want to create a new data channel. After we have created a new data channel, we should ask the remote peer to do the same:

```
function createDataChannel(role) {
    try {
```

When we create a new data channel, we can set up a name of the channel. In the following piece of code, we will use the number of the virtual room to name the channel:

```
        data_channel =
        pc.createDataChannel("datachannel_"+room+role, null);
    } catch (e) {
        console.log('error creating data channel ' + e);
        return;
    }
    initDataChannel();
```

```
    }

    function onIceCandidate(event) {
        if (event.candidate)
            sendMessage({type: 'candidate', label:
            event.candidate.sdpMLineIndex, id:
            event.candidate.sdpMid,
            candidate: event.candidate.candidate});
    };

    function failureCallback(e) {
        console.log("failure callback "+ e.message);
    }

    function doCall() {
```

When we are playing the role of the connection initiator (caller), we create a new data channel. Then, during the connection establishment, the remote peer will be asked to do the same and the data channel connection will be established:

```
        createDataChannel("caller");
        pc.createOffer(setLocalAndSendMessage, failureCallback,
        null);
    };

    function doAnswer() {
        pc.createAnswer(setLocalAndSendMessage, failureCallback,
        null);
    };

    function setLocalAndSendMessage(sessionDescription) {
        pc.setLocalDescription(sessionDescription);
        sendMessage(sessionDescription);
    };
```

To send text messages via the data channel, we need to implement the appropriate function. As you can see in the following code, sending data to the data channel is pretty easy:

```
    function sendDataMessage(data) {
        data_channel.send(data);
    };
```

The following handler is necessary to print the state of the data channel when it is changed:

```
    function onChannelStateChange() {
        console.log('Data channel state is: ' +
        data_channel.readyState);
    }
```

When the remote peer sends us a message via the data channel, we will parse it and call the appropriate function to show the message on the web page:

```
function onReceiveMessageCallback(event) {
    console.log(event);
    try {
        var msg = JSON.parse(event.data);
        if (msg.type === 'chatmessage')
        onPrivateMessageReceived(msg.txt);
    }
    catch (e) {}
};
```

Save the JavaScript file.

Now, start the signaling server and open the HTML file in a web browser—you should see an input field and a button on the page. At the top of the page, you should see a URL to be shared with the remote peer.

On another browser's window, open the sharing link. In the web browser's console, you should see the **Data channel state is open** message. Now, enter something in the input box and click on the **Submit query** button. You should see the message printed on another browser's window.

How it works...

When the application starts, it establishes a connection with the signaling server and gets a virtual room number. Then, another peer starts the application and enters the virtual room. The second peer is the caller. When the peer connection is established, the caller creates a new data channel and another peer receives this event notification. So, both peers get a data channel reference and can use it for data exchanging.

In our example, when the customer enters a message and clicks on the **Submit query** button, we will wrap the message into a JSON object and send it via the data channel. The remote peer gets the JSON object, parses it to the message, and displays it on the page.

There's more...

Using data channels, peers can exchange any kind of data. It can be plain text, for example, or binary data. Moreover, the same data channel can be used to exchange different sorts of data between peers. In this recipe, we used JSON to format messages, and every packet has a `type` field. To send text messages, we used the `chatmessage` type, but you can use your own custom type system to distinguish messages. You can also use something other than JSON. So, data channels are a good tool to exchange data between peers, using a secure and safe direct connection.

- ▶ Please refer to the *Implementing a chat using a signaling server* recipe to learn the other way this feature can be implemented

Implementing a chat using a signaling server

In this recipe, we will cover the process of implementing private, peer-to-peer web chat using signaling server as the middle point. Peers will send chat messages via the signaling server. In the schema represented in the following diagram, you can see the flow:

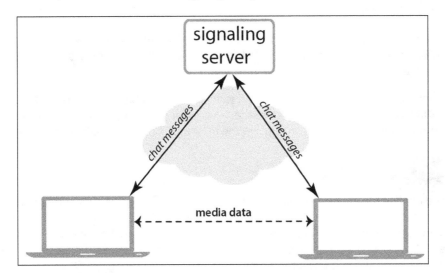

How to do it...

To implement the chat feature via the signaling server, we need to add some methods to the client code with the following steps:

1. We need to add appropriate code to the function that processes the messages from the signaling server:

```
function processSignalingMessage(message) {
        var msg = JSON.parse(message);
        if (msg.type === 'CHATMSG') {
            onChatMsgReceived(msg.value);
        } else if (msg.type === 'offer') {
            pc.setRemoteDescription(new
            RTCSessionDescription(msg));
            doAnswer();
        } else if (msg.type === 'answer') {
```

```
            pc.setRemoteDescription(new
            RTCSessionDescription(msg));
        } else if (msg.type === 'candidate') {
            var candidate = new
            RTCIceCandidate({sdpMLineIndex:msg.label,
            candidate:msg.candidate});
            pc.addIceCandidate(candidate);
        } else if (msg.type === 'GETROOM') {
            room = msg.value;
            onRoomReceived(room);
        } else if (msg.type === 'WRONGROOM') {
            window.location.href = "/";
        }
};
```

2. We will check whether the received message is of the CHATMSG type and if so, we will call the onChatMsgReceived method to process it:

```
function onChatMsgReceived(txt) {
    var chatArea = document.getElementById("chat_div");
    chatArea.innerHTML = chatArea.innerHTML + txt;
    chatArea.scrollTop = chatArea.scrollHeight;
};
```

Here, we will get the chat_div element by its ID and alter its content by adding the chat message received from the remote peer via the signaling server.

3. To send a chat message, we should implement a method like the following:

```
function chatSendMessage(msg) {
        if (!channelReady) return;
        sendMessage({"type" : "CHATMSG", "value" : msg});
};
```

This function checks whether the WebSocket channel is up and sends a chat message to the signaling server using the channel. To use this function, we can use the HTML input tag with the submit button and call it on the submit event.

How it works...

The basic principle of this solution is pretty simple:

▶ One peer sends a text message to the signaling server, marking it as the CHATMSG type

▶ The signaling server retransmits the message to another peer

▶ Another peer gets the message from the signaling server, checks whether it is of the CHATMSG type and if so, shows it to the user

To distinguish chat messages from WebRTC messages, you can use any word to mark the message type. It can be CHATMSG or whatever you prefer.

There's more...

This way of implementing web chat is usually not secure because the data will go via the signaling server and not directly through the peers. Nevertheless, it is suitable for public chat rooms where there can be several people at a time. For private peer-to-peer chats, it is usually better to use WebRTC data channels, and that way it is more secure.

See also

► To implement the chat feature using data channels, follow the *Implementing a chat using data channels* recipe

Configuring and using STUN

Your WebRTC application can work without STUN or TURN servers if all the peers are located in the same plain network. If your application is supposed to work for peers that might be located in different networks, it will definitely need to use at least the STUN server to work.

Getting ready

In this recipe, we will install a STUN server on a Linux box. STUN server can be installed under the other platform as well, but for simplicity, we will consider only the Linux case. So, please prepare a Linux machine.

In this recipe, we will use a very basic and simple STUN server implementation, so you probably will not need to install additional libraries or do some difficult configuration.

STUN needs two IP addresses to work correctly. Thus, when experimenting with your Linux box, take care that the Linux box should have at least two IP addresses that are available for all possible peers (WebRTC clients).

How to do it...

The following set of steps will lead you through the process of configuring and building a STUN service:

1. Download the STUN server from its home page at http://sourceforge.net/projects/stun/.

2. Unpack the archive and go into the STUN server folder:

```
tar -xzf stund-0.97.tgz
cd stund
```

3. Build it with the following command:

```
make
```

The last command will build the server. After that, you can start the STUN server by using the following command:

```
./server -h primary_ip -a secondary_ip
```

Note that instead of `primary_ip` and `secondary_ip`, you should use actual IP addresses that are available on the machine. This software can't detect such network parameters automatically, so you need to set it up explicitly.

 If you want to start the server in the background, add the `-b` option to the preceding command.

Now, when the STUN server is configured and running, we can utilize it in the WebRTC application. When your application wants to create a peer connection object, it uses something like the following code:

```
var pc;
pc = new RTCPeerConnection(configuration);
```

Here, `configuration` is an entity that contains different options for creating peer connection object. To utilize your freshly installed STUN server, you should use something like the following code:

```
var configuration = {
  'iceServers': [
    {
      'url': 'stun:stun.myserver.com:19302'
    } ] }
```

Here we inform the web browser that it can use the STUN server if necessary. Note that you should use the real domain name or IP address of the STUN server. You can also explicitly set the port number as shown in the preceding code, in case it is distinguished from the default value.

How it works...

STUN server can help peers determine their network parameters and thus establish a direct communication channel. If your clients are located behind NAT or firewall, your application should use at least the STUN service to make the direct connection possible. Nevertheless, in many cases that might not be enough, and using TURN might be necessary.

The following diagram might be helpful to you to imagine how the STUN server is located in the whole infrastructure, and how all the components interoperate with each other:

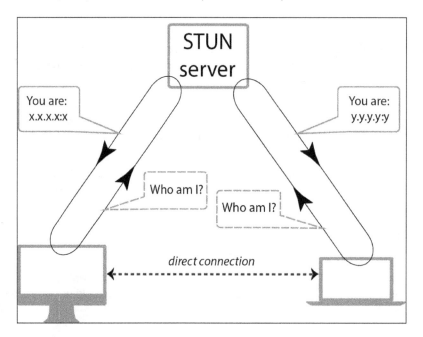

There's more...

As an alternative to this, you can use rfc5766-server—it is a free and open source implementation of both STUN and TURN servers. It also supports many additional features that might be quite useful. You can find it at `https://code.google.com/p/rfc5766-turn-server/`.

See also

►   For details on how STUN works, you can refer to RFC #3489 `http://www.ietf.org/rfc/rfc3489.txt`.

►   In the *Configuring and using TURN* recipe, we will use a TURN server based on the rfc5766-server software. That application can serve as a STUN server as well.

Configuring and using TURN

In most cases, it is enough to use a STUN server to establish a peer-to-peer direct connection. Nevertheless, you will often need to utilize TURN servers—mostly for clients located in big companies (because of firewall policy and tricky NAT) and some specific countries (because of firewalls and access limits).

Getting ready

In this section, we will download, install, and do the basic configuration of a TURN service. Then, we will utilize it in our WebRTC application. A TURN server can be installed under different platforms, although we will cover a Linux box use case only. Thus, for this recipe, you will need a Linux box installed.

For this recipe, we will use rfc5766-turn-server—a free and open source implementation of the TURN and STUN servers. Download its source code from its home page at `https://code.google.com/p/rfc5766-turn-server/`.

How to do it...

First, we will shortly cover the installation and basic configuration of the TURN server. After that, we will learn how to use it in the application.

If you have TURN server already installed, you can skip this section and go directly to the next one.

Installing the TURN server

I assume that you have downloaded rfc5766-server already and unpacked it. So, let's install and configure your own TURN server:

1. Go to the `rfc5766-server` folder with the following command:

   `cd ~/turnserver-4.1.2.1`

2. Build the server:

   `./configure`

   `make`

   `sudo make install`

Note that rfc5766-server needs some libraries that might be not installed on your system—in particular, `libssl-dev`, `libevent-dev`, and `openssl`. You should install the absent libraries to compile the software successfully.

3. After that, you can start the server—it will detect all the network options automatically:

    ```
    turnserver
    ```

    You will see diagnostic messages in the console:

    ```
    0: ==========Discovering relay addresses: =============
    0: Relay address to use: x.x.x.x
    0: Relay address to use: y.y.y.y
    0: Relay address to use: ::1
    0: ========================================================
    0: Total: 3 relay addresses discovered
    0
    0: ========================================================
    ```

 To stop the server, just press *Ctrl* + *C*; you will get back to console.

Now it is time to perform some configuration steps and tune your fresh TURN server for your requirements.

By default, the TURN server doesn't have any configuration file. We need to create this configuration file from the default configuration file supplied with the server:

```
sudo cp /usr/local/etc/turnserver.conf.default /usr/local/etc/turnserver.conf
```

Open the `turnserver.conf` file and edit it according to your requirements. We will not cover all the TURN options here, but just basic configuration items that might be important:

▸   **Listening IP**: This option determines the IP addresses that will be used by the TURN server while operating. By default, this option will do it automatically. Nevertheless, it is a good idea to set the obvious IP addresses you would like the server to use:

    ```
    listening-ip=
    ```

Note that the TURN server needs at least two public IP addresses to operate correctly.

▶ **Relay IP**: In this option, you can explicitly set up IP address that should be used for relay. In other words, if you have two IP addresses, one of them can be `listening-ip` and the second one `relay-ip`.

`relay-ip=`

▶ **Verbosity**: In this option, you can set a level of verbosity. By default, the TURN server will not print extra details on its work, but for debugging and diagnostic purposes, it might be very useful to set the verbose level to normal. For that, you should place the word `verbose` in the configuration file. If you would like to refer to more details, you should write the word with capital V—`Verbose`—so the server will print as much debugging details as possible.

▶ **Anonymous access**: You can enable anonymous access during the development process, if you're sure that your TURN server is protected by network firewall and nobody can use it. Otherwise, you should not enable this option especially on production systems:

`no-auth`

In this recipe, we haven't covered TURN authentication—this topic is covered in *Chapter 2, Supporting Security*.

At this stage, you have your own TURN server with basic configuration, which can be used in WebRTC applications.

Using TURN in WebRTC application

When you create a peer connection object, you usually use some construction like the following one:

```
var pc;
pc = new RTCPeerConnection(configuration);
```

Here, `configuration` is an entity that contains different options to create a peer connection object. To utilize your TURN server, you should use something like the following:

```
var configuration = {
  'iceServers': [
    {
      'url': 'stun:stun.l.google.com:19302'
    },
    {
```

```
        'url': 'turn:turn1.myserver.com:3478?transport=udp',
    },
    {
      'url': 'turn:turn2.myserver.com:3478?transport=tcp',
      'credential': 'superuser',
      'username': 'secretpassword'
    }
  ]
}
```

Here, we will ask the WebRTC API (actually, we will ask the web browser) to use one of three ways when establishing a peer connection:

► Public STUN server provided by Google.

► TURN server with anonymous access. You will use this notation to utilize the TURN server installed and configured in this recipe.

► TURN server with authentication. In *Chapter 2, Supporting Security*, we will cover the topic of security and authentication within the scope of a TURN server. To utilize a server that uses authentication, you should use this notation.

Note that you can ask the web browser to use a UDP or TCP protocol while establishing a peer connection through the TURN server. To do that, set up the transport parameter as shown in the preceding bullet points.

How it works...

In some cases, when clients use NAT and firewalls, it is impossible to establish a peer connection using STUN. This situation often appears when a client is located in a corporative network with a strict policy. In such a case, the only way to establish the connection is to use the TURN server.

The TURN server works as a proxy—all the data between peers (including audio, video, and service data) goes through the TURN server.

The following diagram shows how all the components operate with each other:

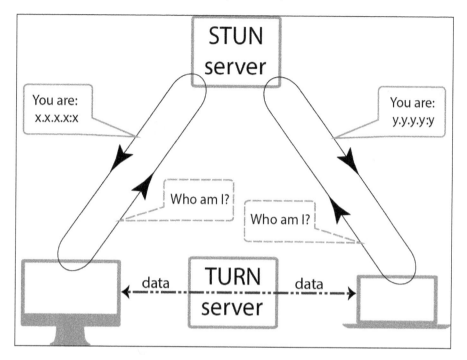

There's more...

In this recipe, we covered only one TURN solution, open source and popular, but there are other solutions in the world that could also be suitable for you:

- **TurnServer**: This is also free and open source. For more information, refer to `http://turnserver.sourceforge.net`.

- **Numb**: This is not software that you can download and install, but a service where you can create an account and get access to a configured TURN server. For more details, refer to `http://numb.viagenie.ca`.

Of course, there are even more different solutions and services available.

See also

- For details on TURN servers, refer to RFC 5766 at `http://tools.ietf.org/html/rfc5766`

- For details regarding STUN (another useful technology with the scope of developing WebRTC-based services), you can also take a look at the *Configuring and using STUN* recipe

2

Supporting Security

In this chapter, we will cover the following topics:

- ▶ Generating a self-signed certificate
- ▶ Configuring a TURN server with authentication
- ▶ Configuring a web server to work over HTTPS
- ▶ Configuring a WebSockets proxy on the web server
- ▶ Configuring a firewall

Introduction

At the time of writing this book, the WebRTC standard was not complete and the technology and its standard were both under active development. Nevertheless, security and safety are very important and mandatory functional requirements that lie at the basis of the WebRTC standard. Basically, your WebRTC application should use only encrypted channels.

In this chapter, we will cover security-related questions. We will talk about security, third-party components, and software you will probably use when developing your WebRTC service. We will talk about configuring secured channels (HTTPS) on web browsers. We will cover the process of creating secure certificates and using them in web servers as well as the TURN service. We will also learn how to implement authentication on a TURN server using the TURN REST API.

We will talk about how WebRTC can work through firewalls and NAT, and learn how to configure a firewall on our server that is serving auxiliary services such as TURN or STUN.

Generating a self-signed certificate

Using encryption is highly recommended (I'd say even mandatory) for WebRTC applications. The technology has good support for security and encryption, so there is no reason to ignore them. In this recipe, we will cover the process of creating self-signed certificates. Such a certificate can be used with a TURN server or with a web server when operating with HTTPS channels.

Typically, a **public key infrastructure** (**PKI**) is a digital signature from a **certificate authority** (**CA**), which attests that a particular PKI is valid and contains correct information. Users or their software then check that the private key used to sign a certificate matches the public key in the CA's certificate. Since CA certificates are often signed by other, *high-ranking* CAs, there must necessarily be a highest-ranking CA, which provides the ultimate attestation authority in that particular PKI scheme.

The highest-ranking CA's certificates are termed as **root certificates**. Clearly, the lack of mistakes or corruption in the issuance of such certificates is critical to the operation of its associated PKI; they should be, and generally are, issued with great care.

A self-signed security certificate is a certificate that is signed by the same entity whose identity it certifies. Such a certificate can be used for developing purposes and can be generated by anybody.

You can find more details on PKI at `http://en.wikipedia.org/wiki/Public_key_infrastructure`.

 Self-signed certificates can be used for development, but you should issue trusted certificates for production systems.

All communication channels in a WebRTC application should be using encryption: client-to-client, client-to-server, or any other kind of channels you might be using. Some WebRTC features (such as screen sharing) will not work without encryption, even in a development environment.

Getting ready

In this recipe, we will use the OpenSSL toolset.

OpenSSL is an open source multiplatform toolkit that implements **Secure Sockets Layer** (**SSL**) and **Transport Layer Security** (**TLS**) protocols and provides a general purpose full-strength cryptography library. Many computer software use OpenSSL for supporting encryption and security.

You can find more details on this product at its home page, `https://www.openssl.org`.

Often, it is installed by default on Unix-like systems, but it is not supplied with Windows installations. To check whether your system has OpenSSL installed, you can use the following console command:

```
openssl version -a
```

On my Mac, it produces the following message:

```
~ >>> openssl version -a
OpenSSL 0.9.8za 5 Jun 2014
built on: Sep  9 2014
platform: darwin64-x86_64-llvm
options:  bn(64,64) md2(int) rc4(ptr,char) des(idx,cisc,16,int) blowfish(idx)
compiler: -arch x86_64 -fmessage-length=0 -pipe -Wno-trigraphs -fpascal-strings -fasm-blocks
OPENSSL_NO_IDEA -DOPENSSL_PIC -DOPENSSL_THREADS -DZLIB -mmacosx-version-min=10.6
OPENSSLDIR: "/System/Library/OpenSSL"
~ >>>
```

If you see something similar, then you have OpenSSL installed on your system. If not, you need to install the tool.

 For more information on how to install OpenSSL, please refer to its official home page at `http://www.openssl.org`.

How to do it...

In this section, we will generate public and private security certificate keys with the following steps:

1. First, generate a temporary server password key:

   ```
   openssl genrsa -des3 -passout pass:x -out server.pass.key 2048
   ```

   You will see something like the following screenshot:

```
~/temp >>> openssl genrsa -des3 -passout pass:x -out server.pass.key 2048
Generating RSA private key, 2048 bit long modulus
........+++
..................................................+++
e is 65537 (0x10001)
~/temp >>>
```

2. Using the server password key, generate a server private key:

   ```
   openssl rsa -passin pass:x -in server.pass.key -out server.key
   ```

   You will see the following output:

   ```
   writing RSA key
   ```

3. We don't need the server password key, so we can remove it:

   ```
   rm -rf server.pass.key
   ```

4. Generate a certificate signing request:

   ```
   openssl req -new -key server.key -out server.csr
   ```

5. This will ask you additional questions about the company the certificate is being created for—you can use fictional data. It will also prompt you for a password as shown in the following screenshot—for simplicity, you can just press return:

```
                                          2. mc [andreysergienko@192.168.1.50]
~/temp >>> openssl req -new -key server.key -out server.csr
You are about to be asked to enter information that will be incorporated
into your certificate request.
What you are about to enter is what is called a Distinguished Name or a DN.
There are quite a few fields but you can leave some blank
For some fields there will be a default value,
If you enter '.', the field will be left blank.
-----
Country Name (2 letter code) [AU]:
State or Province Name (full name) [Some-State]:New-York
Locality Name (eg, city) []:New-York
Organization Name (eg, company) [Internet Widgits Pty Ltd]:MyOrganization
Organizational Unit Name (eg, section) []:UnitOne
Common Name (e.g. server FQDN or YOUR name) []:example.com
Email Address []:admin@example.com

Please enter the following 'extra' attributes
to be sent with your certificate request
A challenge password []:
An optional company name []:
~/temp >>> 
```

6. Generate the certificate:

   ```
   openssl x509 -req -days 365 -in server.csr -signkey server.key
   -out server.crt
   ```

You will see the following output:

```
2. mc [andreysergienko@192.168.1.50]:~/dev/webrtcblueprints.com (mc)
~/temp >>> openssl x509 -req -days 365 -in server.csr -signkey server.key -out server.crt
Signature ok
subject=/C=AU/ST=New-York/L=New-York/O=MyOrganization/OU=UnitOne/CN=example.com/emailAddress=admin@example.com
Getting Private key
~/temp >>>
```

Now you have two files, `server.crt` (the certificate) and `server.key` (the certificate's private key), which can be used with your web server (operating over HTTPS) or TURN server.

How it works...

By using an OpenSSL tool, we generated a new self-signed security certificate that can be used with a web server or a TURN server that is serving our WebRTC application.

 Kindly note that we generated the certificate in PEM format. For some software, it might be necessary to convert it to other formats.

Though the certificate implements full encryption, your website visitors will see a browser warning indicating that **The certificate should not be trusted!**.

If a self-signed certificate has been used to create a WebSocket server, then your web browser will fail when trying to establish a connection to the server and will not show any warning. To solve such a case, you can configure a web server to be secured, but leave the WebSocket server unsecured; then, you should configure a WebSocket proxy on the web server. Thus, the client will communicate with the WebSocket server not directly, but through the web server using a secured channel. Please refer to the *Configuring a WebSockets proxy on the web server* recipe.

So, use self-signed certificates for developing only.

There's more...

For production systems, you should use trusted certificates emitted by such trusted centers such as Verisign, Thawte, or others.

You can also start with a free-of-charge but trusted certificate from StartSSL. For more details, refer to `http://www.startssl.com`.

If you have a Windows box, you can use the SelfSSL.exe tool to create a self-signed certificate. This tool is part of **Internet Information Services (IIS) Resource Kit Tools** that can be found at `http://www.microsoft.com/en-gb/download/details.aspx?id=17275`.

You can also use online tools to create a self-signed certificate, for example, this one at `http://www.selfsignedcertificate.com`.

See also

> ▶ You can find more details on how to use certificates in the *Configuring a TURN server with authentication* and *Configuring a web server to work over HTTPS* recipes.

Configuring a TURN server with authentication

STUN servers don't support authentication, but on the other hand, TURN servers do. Moreover, if you maintain a TURN server, it has to support authentication and prohibit anonymous access. When using a TURN service, all the traffic from one peer to another goes through the TURN server. If anyone had anonymous access to such a server, they could very quickly utilize the server's resources and traffic limits.

In this recipe, we are going to go through a TURN authentication task.

Getting ready

First of all, we need to download and install a TURN server. There are several implementations, and in this recipe, we will consider using **rfc5766-turn-server**.

This software is multiplatform and can be used on Unix-like systems and on Windows systems as well. Nevertheless, to keep it simple, in this recipe, we will cover a Linux-based case only.

Download the source code from the TURN server home page at `https://code.google.com/p/rfc5766-turn-server/`.

To install the software, you might need other additional packages to be installed first:

> ▶ `mysqlclient-dev`
>
> ▶ `libevent`
>
> ▶ `libmysqlclient-dev`
>
> ▶ `libevent-dev`
>
> ▶ `libssl-dev`

Please use your package installation tool to install necessary packets.

The package list might vary for different Linux distributions.

Unpack the downloaded TURN server package into a new folder, go to it, and then compile and install the software using the following commands:

```
./configure
```

```
make
```

```
sudo make install
```

If you didn't change the installation prefix, the configuration file will be placed at `/usr/local/etc/turnserver.conf`.

Now, we need to edit this file, changing the necessary options. We will not cover all the configuration options, but just the ones that are necessary to achieve our goal:

1. First, ensure the support for encrypted transport:

   ```
   tls-listening-port=5349
   ```

2. Switch the verbose mode on:

   ```
   verbose
   ```

 You don't want verbose enabled on a production system, but it is very useful for debug purposes. I'd recommend you keep it enabled during the developing/debugging process, and then disable it when you deploy your application to the production system.

3. Enable a long-term credential mechanism—a REST API can be used with long-term credentials only:

   ```
   lt-cred-mech
   ```

4. Comment out the short-term credential mechanism option:

   ```
   #st-cred-mech
   ```

5. Enable the REST API:

   ```
   use-auth-secret
   ```

6. Determine the static authentication secret—the client will use this value when calculating the temporary password for accessing the TURN server:

   ```
   static-auth-secret=<SuperSecretKey>
   ```

7. Set up the realm (usually, the company's website domain name):

```
realm=mycompany.org
```

8. Set up the generated security certificate:

```
cert=/usr/local/etc/turn_server_cert.pem
```

9. Set up the certificate private key:

```
pkey=/usr/local/etc/turn_server_pkey.pem
```

10. Set up the security certificate key password. This option is important if you use a certificate protected by a password. If the key is password-less, then leave this option commented out:

```
#pkey-pwd=
```

11. Using the enable console feature, you can connect to the TURN server console and control the server, or just get some statistics. It is very useful for debugging:

```
cli-ip=127.0.0.1
cli-port=5766
```

12. Set up the console password:

```
cli-password=<you-cli-password>
```

How to do it...

Now we have a TURN server installed and configured. Next, we need to make appropriate changes on the client-side code (that will be executed by the web browser) and on the server-side code.

What is important for this feature is that your web application should have a user authentication mechanism implemented. The application should have a private area where only authorized users can get access.

The certain implementation depends on the platform/framework you use for developing the application.

Implementing the client-side code

The general flow to implement the client-side code is as follows:

1. The application has some kind of authorization form with a login and password. The web page with the WebRTC feature should be hidden behind the login page and access should be restricted to authorized users only.

2. After the user enters correct credentials and has been authorized, he/she can have access to a private area.

3. When a user is authorized, he/she should be forwarded to the private area where the WebRTC interactive page is placed.

4. The interactive page that a user gets from the web server should contain correct credentials for accessing the TURN server. These credentials are calculated by the web server and are then sent to the authorized client.

The following fragment is an example of what an authorized client should get from the web server:

```
var iceServers = [
    {
        'url' : 'turn:turn1.website.com',
        'credential' : 'dejwjhkuyui4BUHiebdiejbi',
        'username' : 'secretuser'
    }
];
```

Here, the `credential` field is the temporary password calculated on the web server that can be used to access the TURN server. Only authorized users can get it—username is also used while calculating the password (refer to the *Implementing the server-side code* section of this recipe).

The client should use this data when accessing the TURN server (pseudo code):

```
pc_config = {"iceServers": iceServers};
var pc = RTCPeerConnection(pc_config, pc_constraints);
```

Implementing the server-side code

Server-side code can be implemented using any language and technology you like. If you are a Java programmer, the easiest way would be to use Java Spring or Play Framework.

The web server should provide the following flow for implementing server-side code:

1. Authenticated users should access the WebRTC interactive web page only.

2. When a user is authenticated using the login page, they should be forwarded to the private area (the WebRTC interactive page).

3. During the authentication process, the web server should store the user login.

4. The web server should calculate the TURN temporary access password using the following formula:

```
base64(hmac(secret key, username))
```

Here you can see the following:

- ❏ `secret key`: This is the static-auth-key option from the TURN server configuration (refer to the *Getting ready* section of this recipe)

- ❏ `username`: This is the username the web server gets from the user during the authentication

- ❏ `hmac`: This is the hash function of `secret key` and `username`

- ❏ `base64`: This function implements the `base64` encoding algorithm, and we apply it to the result of the `hmac` function

5. After the temporary password is calculated, it should be sent to the client.

How it works...

In this recipe, we will utilize the **TURN REST API**. The main goal of this API is to provide a mechanism that will enable dynamic temporary passwords, which can be used with TURN servers when authenticating.

The general TURN authentication flow is as follows:

- ▶ The client (a web browser) sends a request to the application (that is working on the server side) asking for TURN credentials. Optionally, the request can also include the username.

- ▶ The application responds with a TURN URL, username, and password.

The client then uses these credentials for further authentication on the TURN server. The application then replies with the following data:

- ▶ **Username**: This is the TURN username that the client has to use when authenticating. This name is a colon-delimited combination of the expiration timestamp and username parameter from the client's original request. If the username is not specified, the server can use any other value here.

- ▶ **Password**: This is the TURN password that the client has to use when authenticating. This value is calculated by the server using the following algorithm: `base64(hmac(secret key, username))`. The TURN server and the server application both share the same secret key. So, the TURN server will do the same calculations and will compare them to the credentials received from the client.

 Kindly note that credentials are temporary (time limited).

▸ **TTL**: This value represents the time-to-live parameter. It is optional and we won't use this field on our application.

▸ **URIs**: This field represents an array of URLs of the TURN server(s) available. In our case, we will send just one URL to our own TURN server.

There's more...

This feature is not a part of the final standard yet, so in the future, some aspects of this recipe might need to be improved.

See also

▸ Refer to the *Generating a self-signed certificate* recipe on how to create self-signed certificates.

▸ Take a look at the TURN server's REST API standard draft at `http://tools.ietf.org/html/draft-uberti-rtcweb-turn-rest`.

▸ The PDF from the rfc5766-trun-server documentation can be useful. For more details, refer to `https://rfc5766-turn-server.googlecode.com/svn/docs/TURNServerRESTAPI.pdf`.

Configuring a web server to work over HTTPS

In this recipe, we will cover how to configure a secured layer (HTTPS) on a web server. As far as encryption and security are mandatory for WebRTC, HTTPS is an important part of the whole application's security and safety.

Getting ready

We will cover the three most popular web servers: Nginx, Apache HTTP Server, and IIS from Microsoft. We will not cover the installation procedure, so you should have the web server you wish to use installed and properly configured.

How to do it...

What we need to do is to edit the web server's configuration to switch on using HTTPS. Before we make the configuration changes, you need to have the generated security certificate. Usually, it is two files: a certificate and certificate key. But it is possible to join these two files into just one. In this recipe, we will consider the first option, with two files.

These certificate files (`server.crt` and `server.key`) can be trusted SSL certificates or they can be self-signed certificates.

Configuring Nginx

You should edit the website's configuration file—usually, it is located under `/etc/nginx/sites-enabled/website.com`:

1. The following configuration fragment shows important changes that you should make:

    ```
    server {
    ```

2. We will ask the web server to listen on port `443` (default port for HTTPS) and use SSL:

    ```
    listen              443 ssl;
    ```

3. You should also set up the website's name—as you would do for a non-secured website:

    ```
    server_name         www.example.com;
    ```

4. For a secured website, we need to set up the SSL certificate and SSL certificate key. Technically, they're just two files generated in a specific way:

    ```
    ssl_certificate     /etc/nginx/ssl/certs/server.crt;
    ssl_certificate_key /etc/nginx/ssl/private/server.key;
    }
    ```

     A good practice is to keep `.crt` and `.key` files in different folders, as you can see in the preceding code. So don't forget to copy both files of your security certificate to proper places. Create appropriate folders if necessary.

5. You will need to reload the web server after these changes. For Ubuntu, this can be done using the following command:

    ```
    sudo service nginx reload
    ```

6. Alternatively, you can restart the whole web server using the following command:

    ```
    sudo service nginx restart
    ```

Configuring Apache

You should edit the website's configuration file—usually, it can be found under `/etc/apache2/sites-available/website.conf`.

We will not cover all the configuration files but will consider relevant changes:

1. Add the option to make the Apache web server listen on the HTTPS default port `NameVirtualHost *:443`.

2. Make necessary changes in the appropriate `VirtualHost` section:

    ```
    <VirtualHost *:443>
    ServerAdmin webmaster@website.com
    DocumentRoot /var/www/website.com
    ServerName www.website.com
    DirectoryIndex index.php
    ErrorLog /var/log/apache2/vhost1-error.log
    CustomLog /var/log/apache2/vhost1-access.log combined
    SSLEngine On
    SSLCertificateFile /etc/apache2/ssl/server.crt
    SSLCertificateKeyFile /etc/apache2/ssl/server.key
    <Location />
    SSLRequireSSL On
    SSLVerifyClient optional
    SSLVerifyDepth 1
    SSLOptions +StdEnvVars +StrictRequire
    </Location>
    </VirtualHost>
    ```

3. After you've made these changes, you need to restart the Apache HTTP Server.

Configuring IIS

In this section, we will cover how to configure IIS to use the SSL certificate:

1. Log on to the web server computer as an administrator.

2. Click on **Start**, point to settings, and then click on **Control Panel**.

3. Double-click on **Administrative Tools**, and then double-click on **Internet Services Manager**.

4. Select the website from the list of different served sites in the left pane.

5. Right-click on the website on which you want to configure SSL, and then click on **Properties**.

6. Click on the **Directory Security** tab.

7. Click on **Edit** and then on **Require secure-channel (SSL)**.

8. Click on **Require 128-bit encryption** to configure 128-bit (instead of 40-bit) encryption support.

9. To allow users to connect without supplying their own certificate, click on **Ignore client certificates**:

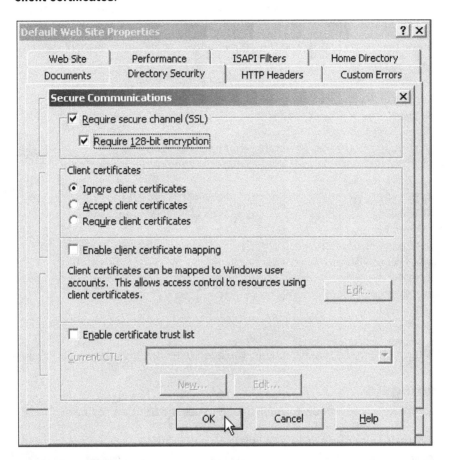

There's more...

The configuration process might vary depending on the certain web server version you use. Please refer to the appropriate vendor's documentation provided with the web server you use for specific details.

See also

Refer to the *Generating a self-signed certificate* recipe for details on how to create self-signed certificates. There you can also find additional information on where to start if you would like to get a trusted certificate to use in production.

Configuring a WebSockets proxy on the web server

WebSockets is a new protocol that enables active messaging from server to client. It is supported by all modern web browsers. This protocol is implemented on top of HTTP and can be easily served by most popular web servers. It can also be served over secured channels, such as HTTPS. Because of WebSockets' advantages, people often choose this protocol for their client-server projects. In WebRTC-based applications, WebSockets usually serves as a transport protocol for signaling server implementation.

Configuring a WebSockets proxy on a web server can be very useful if you have used WebSockets as a transport layer for communicating with the signaling server. For some cases, it might even be mandatory.

Getting ready

Configuring this feature requires making changes in the configuration files of the web server. We will not cover the entire web server's installation and configuration process, so you need to have the web server up and running.

How to do it...

We will make necessary configuration changes to the web server to achieve the goal.

Configuring Nginx

The website's configuration files are usually located under the `/etc/nginx/sites-enabled` folder:

1. The following piece of the website configuration file shows the WebSockets proxy settings:

   ```
   location /websocket {
   ```

2. Here we will set the local service that will be serving WebSocket requests for the web server:

   ```
   proxy_pass http://localhost:16384;
   ```

3. Indicate that we will work with HTTP protocol version 1.1—WebSockets is not supported on lower HTTP versions:

   ```
   proxy_http_version 1.1;
   proxy_set_header Upgrade $http_upgrade;
   proxy_set_header Connection "upgrade";
   proxy_redirect off;
   ```

4. Here we can set additional options asking the web server to send useful details about connected clients in the HTTP headers:

```
proxy_set_header    Host                $host;
proxy_set_header    X-Real-IP           $remote_addr;
proxy_set_header    X-Forwarded-For
$proxy_add_x_forwarded_for;
}
```

5. You will need to restart Nginx or reload its configuration after you've made these changes.

 You need a separate location section for every WebSocket URL that you need to proxy via the web server.

Configuring Apache

Apache doesn't support this feature from scratch (at least, for versions >= 2.4). Nevertheless, there are some third-party modules that can help us with this. In this recipe, we will use the `apache-websocket` module, available at `https://github.com/disconnect/apache-websocket`:

1. The following configuration fragment shows how to use the module:

```
<IfModule mod_websocket.c>
    <Location /websocket>
        SetHandler websocket-handler
        WebSocketHandler
        /usr/lib/apache2/modules/mod_websocket_tcp_proxy.so
        tcp_proxy_init
        WebSocketTcpProxyBase64 on
        WebSocketTcpProxyHost localhost
        WebSocketTcpProxyPort 16384
        WebSocketTcpProxyProtocol base64
    </Location>
</IfModule>
```

2. On a default Apache installation, you might want to change your request read timeout option:

```
<IfModule reqtimeout_module>
  RequestReadTimeout body=300,minrate=1
</IfModule>
```

3. You can track native support of this feature in Apache using `https://issues.apache.org/bugzilla/show_bug.cgi?id=47485`.

Configuring IIS

The WebSocket proxy feature is available in IIS version 8 and is not supported in older IIS versions.

You should install **Application Request Routing** (**ARR**) 3.0 or a newer version. This is a proxy-based routing module that serves to forward HTTP requests to content servers.

According to Microsoft's recommendations, ARR should be installed using the **Web Platform Installer** (**WebPI**) module.

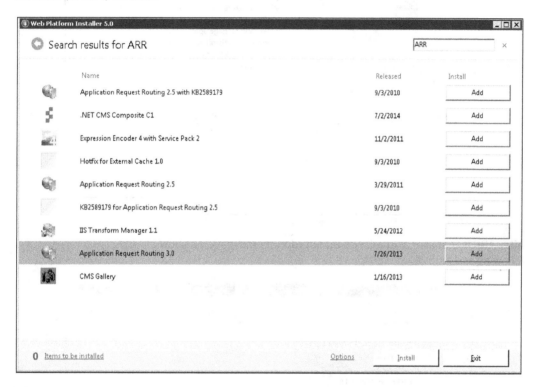

Choose **Application Request Routing 3.0** as depicted in the preceding screenshot, and click on the **Add** button and then click on **Install**. During the installation process, you will see the screen shown in the following screenshot:

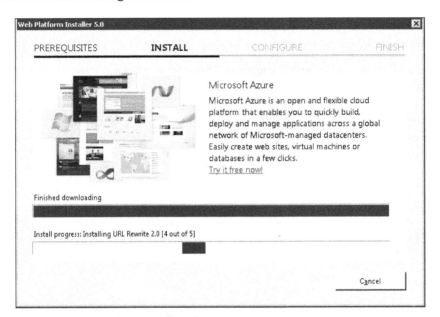

After ARR is installed, you will see an appropriate message as shown in the following screenshot:

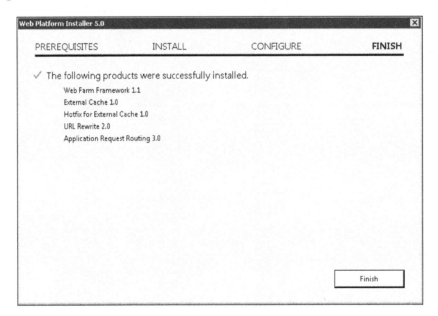

Now, when ARR is installed successfully, you should install the WebSockets features on IIS using the **Server Manager** component and its **Manage** and **Add Roles and Features** menus, as shown in the following screenshot:

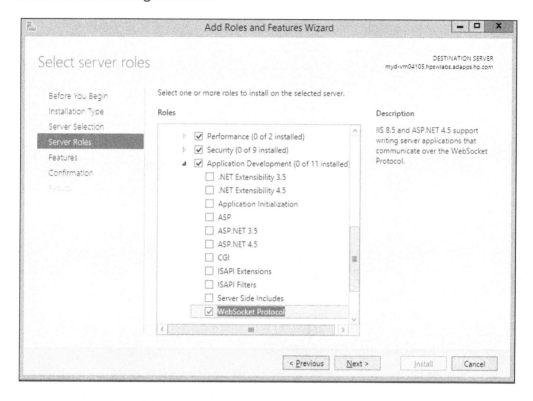

Once the installation is complete, ARR will handle WebSockets requests appropriately.

How it works...

Using secure channels is mandatory for a WebRTC application. In our recipes, we have used WebSockets as a transport protocol to communicate with the signaling server. The main goal of using a WebSockets proxy is to hide the WebSockets service (signaling server) behind the web server, which is serving over HTTPS (secured layer). In such a case, we don't need to configure HTTPS on the signaling server itself.

The following diagram depicts the way this works. When the client (web browser) makes a request to the signaling server using WebSockets, it doesn't make the request to the signaling server directly, but to the web server (using a secured channel, HTTPS). Then, the web server forwards this request to the signaling server (using the usual, non-secured layer), and then it forwards the response back to the client (the web browser).

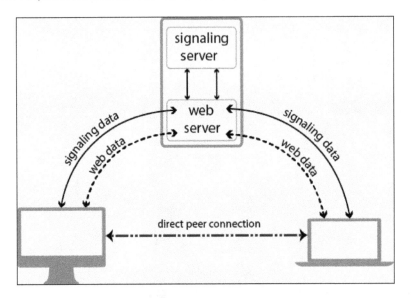

The benefits of using such a solution are as follows:

▶ You don't need to open a listening port of the signaling server to the external world

▶ You need to configure a secured layer for the web server only, and no need to configure it for the signaling server

▶ On the client side, you can use the same domain and port

 These benefits are relevant if you have a web server and signaling server both installed on the same machine.

There's more...

The configuration process might vary depending on the certain web server version you use. WebSockets is a young technology and is not supported by old web servers. The web server should support HTTP 1.1 to be able to support WebSockets and the WebSockets proxy. So, you should use the newest web server version.

You're not limited to using WebSockets for the signaling server transport protocol. This is just a particular case. You can use any transport you like for this purpose. So, if you prefer to use something different rather than WebSockets, this proxy feature might not be relevant for you, or you might have to use other solutions to make your protocol of choice secure and safe.

See also

▶ In this recipe, we built the signaling server behind the web server to utilize its HTTPS (secured layer). For more details on how to configure HTTPS for web servers, please refer to the *Configuring a web server to work over HTTPS* recipe.

Configuring a firewall

If you develop a WebRTC application and maintain your own infrastructure (STUN/TURN servers, web servers), then a properly configured firewall is very important for you. Usually, every server has a network firewall configured and running. Misconfigured firewalls can block services and cause side effects. With WebRTC, a misconfigured firewall can lead the application to **DoS** (**denial of service**) or make some parts of it unworkable; for example, you can hear audio but can't see video.

In this recipe, we will cover basic information that might help you to configure a network firewall properly.

Getting ready

There are many firewall implementations, and it is impossible to cover all of them. So here, we will mostly talk about recommendations rather than practical commands and codes.

Find which firewall is used on your system. On Windows, it is a built-in firewall. On Linux systems, you often have iptables. On BSD systems, it can be pf or ipfw. Mac systems usually use tools from the BSD family. Your system might even be using some kind of third-party tool, so you should refer to the relevant documentation of the firewall tool that is used on your system.

How to do it...

It is worth considering how to configure a firewall in the scope of server side and client side separately.

Configuring a firewall on a server

If you have your own server(s) for all your WebRTC application components (STUN, TURN, web server, any other kind of network services), it is worth knowing which ports and protocols can be used by these components to create an appropriate networking policy. Otherwise, your application might fail to access these services.

Default port numbers for known services relevant for WebRTC applications are as follows:

▸ STUN/TURN: Ports 3478 and 5349, UDP and TCP. The second port is used for TLS.

▸ Web: TCP port 80 for HTTP and TCP port 443 for HTTPS.

▸ Signaling server: This depends on the technology and protocol you use. Using this is a good idea if you can hide the signaling server behind the web server so that the signaling server can listen on localhost only, and not listen to the external world.

▸ If you're using TURN, your server should have two IP addresses—keep this fact in mind when configuring a firewall.

▸ All the preceding ports should be opened and accessible to the external world.

Configuring a firewall on a client

Of course, you can't control a firewall on the user's side. Nevertheless, the following details could help you while debugging or problem solving.

WebRTC has great built-in mechanisms and features to handle firewalls and **Network Address Translation** (**NAT**). It can utilize **Interactive Connectivity Establishment** (**ICE**), which it supports via TURN and using STUN services.

The following screenshot shows two cases:

▸ The communication process between clients and the signaling server

▸ Direct communication between peers after signalization

Chapter 2

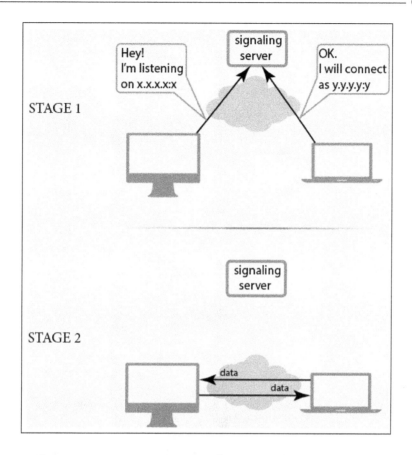

Signalization is necessary when peers are located on different networks. The signaling server should be known and accessible to all peers, and then they can exchange data with each other via network data using the signaling server, and then establish direct connection.

However, if a peer is located behind a NAT/firewall, then this will not work—peers have no way to know their own external network addresses, so establishing the direct connection is problematic.

The following screenshot shows such a case:

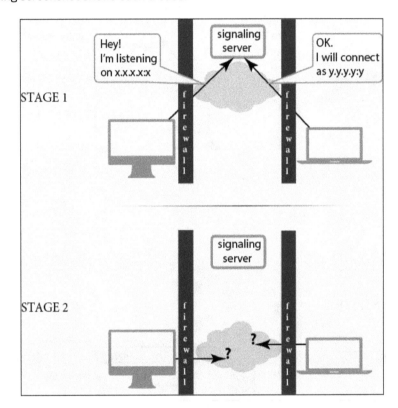

In this case, before peers can establish a direct connection, a STUN service should be used by peers to detect their network parameters. STUN allows peers to know about their external IP addresses. Nevertheless, in many cases it will not work—many users are located behind a NAT and firewall.

If STUN didn't help, the only solution that might solve the issue is a TURN server. In this case, all the data between peers will go through the TURN server—it will proxy all media and other data that peers will transfer to each other. In other words, there will not be a direct peer-to-peer connection established, and all communication will be done via the TURN server in the middle.

This is why you will probably want to have your own TURN server—many business clients have very strong network policies and very complex firewall/NAT configurations, so simple solutions will just not work in their cases.

 If you develop your web application or service using WebRTC, consider installing your own STUN and TURN servers at the very beginning.

The schema in the following diagram depicts the data flow while using a TURN server:

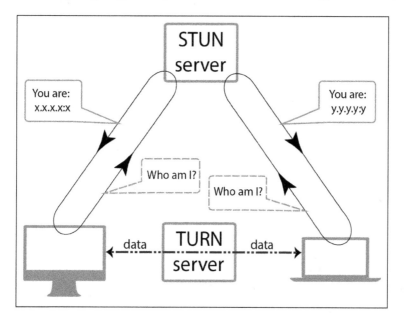

See also

▶ Take a look at the *Configuring a WebSockets proxy on the web server* recipe for details on how to hide the signaling server (the case using WebSockets as the transport layer) behind the web server

3
Integrating WebRTC

In this chapter, we will cover the following topics:

- ▶ Integrating WebRTC with Asterisk
- ▶ Integrating WebRTC with FreeSWITCH
- ▶ Making calls from a web page
- ▶ Integration of WebRTC with web cameras

Introduction

This chapter is fully dedicated to the topic of integrating WebRTC with the rest of the world—other components, technologies, and services.

You will find recipes on integration of WebRTC with VoIP platforms (Asterisk and FreeSWITCH), and will learn how to implement a simple solution in the *Making calls from a web page* recipe using WebRTC and SIP. We will also cover the integration of WebRTC with web cameras.

In this chapter, we will not write code, but will install and configure third-party applications and libraries, connecting them with each other in order to achieve the goal. Most of software that we have covered is cross-platform, but to simplify the task, we will cover Linux-based installations only. So, for most of recipes, you will need to have a prepared Linux machine. Since we will consider simple cases, it will not need many resources, so if you don't have a ready-to-use Linux box, you can use some special software for creating virtual Linux machine to work on the recipes. It can be VMware, VirtualBox, or any other solution you like. You can use any Linux distribution for these purposes; I personally used Ubuntu while working on this book.

Some commands or system paths might be different for different Linux distributions.

The recipes of this chapter don't cover all the completed solutions from scratch, and cover specific questions only. So, it is assumed that you have basic knowledge of using the Linux command line and have basic experience of installing and configuring Linux software.

Integrating WebRTC with Asterisk

In this recipe, we will cover the integration of WebRTC with Asterisk—an open source platform used to build communications applications. Asterisk turns an ordinary computer into a communications server. Asterisk powers IP PBX systems, VoIP gateways, conference servers, and other custom solutions. It is used worldwide by small and large businesses, call centers, carriers, and government agencies.

Asterisk-based telephony solutions offer a rich and flexible feature set. Asterisk offers both classic PBX functionality and advanced features, and interoperates with traditional standards-based telephony systems and Voice over IP systems. Asterisk offers the advanced features that are often associated with large, high-end (and high cost) proprietary PBXs.

Getting ready

In this recipe, we will work under Linux. So, prepare a Linux box. We also will use tools such as Git and SVN—install them if they're not installed yet on your machine.

You might wish to install FreePBX to make your life easier when configuring Asterisk. This software can be found on its home page at `http://www.freepbx.org`.

I assume that you have some experience with installing and configuring Linux software. If not, you can refer to a help page on Linux basics, for example, `http://manuals. bioinformatics.ucr.edu/home/linux-basics`.

How to do it...

During this recipe, we will install and configure a set of applications and build a service by integrating these applications with each other. We will not cover all the installation and configuration steps from scratch, but will cover specific steps only that might be relevant in to this recipe.

Installing libSRTP

Before we compile and install Asterisk, we need to install libSRTP—a software library that provides an **SRTP (Secure Real-time Transport Protocol)** implementation. Asterisk should support SRTP for integrating with a WebRTC application. The support of this protocol is necessary because WebRTC uses secured channels to build communication between peers. We install libSRTP with the following steps:

1. Create a directory `~/src/libsrtp` and go to it.

2. Download `libsrtp` to the folder from the library's home page, `http://sourceforge.net/projects/srtp/files/`.

3. Unpack the downloaded archive and go into the `srtp` folder.

4. Compile the library:

   ```
   ./configure CFLAGS=-fPIC
   make
   sudo make install
   ```

At this point, we have compiled and installed the libSRTP library that will be used when building and installing Asterisk.

Installing Asterisk

In this recipe, we will install Asterisk 11.5; perform the following steps to do so:

1. Download Asterisk from the home page, `http://www.asterisk.org`.

2. Unpack the archive and go into the Asterisk source code folder.

3. Configure Asterisk as follows:

   ```
   ./configure --with-crypto --with-ssl --with-srtp=/usr/local/lib
   contrib/scripts/get_mp3_source.sh
   make menuselect.makeopts
   menuselect/menuselect --enable format_mp3 --enable res_config_
   mysql --enable app_mysql --enable app_saycountpl --enable cdr_
   mysql --enable EXTRA-SOUNDS-EN-GSM
   ```

 Particular configuration options given in the preceding code can vary depending on your specific case. For example, you might be not using MySQL but some other database. In newest versions of Asterisk, `app_saycountpl` is replaced with `app_saycounted`.

4. Build Asterisk as follows:

   ```
   make
   make install
   ```

Now we have compiled and installed Asterisk, we can configure the software with the following steps:

1. Edit `/etc/asterisk/sip.conf` and change the **General** section:

    ```
    udpbindaddr=0.0.0.0:5060
    realm=<your_server_IP >
    transport=udp,ws
    ```

2. Edit `/etc/asterisk/rtp.conf` to enable STUN and ICE:

    ```
    icesupport=yes
    stunaddr=<IP_of_your_STUN_server>
    ```

 If you didn't install your own STUN server yet, you can use the public STUN service from Google at `stun.1.google.com:19302`.

3. Edit `/etc/asterisk/http.conf` and enable an HTTP service:

    ```
    [general]
    enabled=yes
    bindaddr=0.0.0.0
    bindport=8088
    ```

4. Edit `/etc/asterisk/sip.conf` and create a SIP account:

    ```
    [8000]
    secret=SuperS3cret
    context=from-internal
    host=dynamic
    trustrpid=yes
    sendrpid=no
    type=friend
    qualify=yes
    qualifyfreq=600
    transport=udp,ws
    encryption=yes
    dial=SIP/8000
    callerid=John Dow <8001>
    callcounter=yes
    avpf=yes
    icesupport=yes
    directmedia=no
    ```

    You can find additional details on Asterisk configuration options at `http://www.voip-info.org/wiki/`.

5. Now that the configuration is finished, restart Asterisk.

How it works...

The whole schema of interoperation between all the components can be found in the following diagram (taken from the sipML5 library's home page):

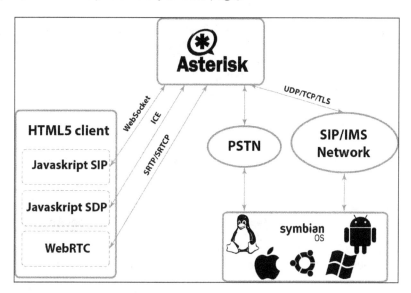

As you can see, **HTML5 client** can interact with a VoIP platform using WebRTC and a SIP module (**JavaScript SIP** in the diagram).

There's more...

There is an opinion that Asterisk is not the best choice. It is perhaps the oldest and most mature solution in the field. Nevertheless, many people found it buggy and unstable in some cases. In particular, WebRTC was not supported by many until the previous versions.

So, if you are looking for alternatives, it might be a good idea to try other solutions such as FreeSWITCH. Its home page can be found at `http://www.freeswitch.org`.

See also

▸ For an alternative solution, using other VoIP software, refer to the *Integrating WebRTC with FreeSWITCH* recipe

▸ In the *Making calls from a web page* recipe, we will cover how to make calls from web pages using WebRTC and a VoIP platform integration

Integrating WebRTC with FreeSWITCH

In this recipe, we will cover the integration of WebRTC with FreeSWITCH—an open source platform used to make VoIP communication services.

FreeSWITCH is a scalable open source cross-platform telephony platform designed to route and interconnect popular communication protocols using audio, video, text, or any other form of media. It was created in 2006 to fill the void left by proprietary commercial solutions. FreeSWITCH also provides a stable telephony platform on which many telephony applications can be developed using a wide range of free tools.

Getting ready

In this recipe, we will work under Linux as well. So, you need a Linux box to be prepared.

It is possible to install FreeSWITCH under Windows, but we don't cover this use case in the recipe. If you need a Windows installation, please refer to the official documentation at `http://wiki.freeswitch.org/wiki/Installation_for_Windows`.

During the work, we will also use tools such as Git and SVN—install them if they're not installed yet on your machine. I assume that you have some experience with installing and configuring Linux software.

How to do it...

During this recipe, we will install and configure a set of applications and build a service by integrating these applications with each other. We will not cover all the installation and configuration steps from scratch, but will only cover the specific steps that might be relevant to this recipe.

Installing FreeSWITCH

FreeSWITCH can be installed from precompiled binary packages or from source code. The first way is easer, but the vendor recommends the second one. We install FreeSWITCH with the following steps:

1. Install the necessary packages for your system:

    ```
    apt-get install autoconf automake devscripts gawk g++ git-core
    libjpeg-dev libncurses5-dev libtool make python-dev gawk pkg-
    config libtiff5-dev libperl-dev libgdbm-dev libdb-dev gettext
    libssl-dev libcurl4-openssl-dev libpcre3-dev libspeex-dev
    libspeexdsp-dev libsqlite3-dev libedit-dev libldns-dev libpq-dev
    ```

2. Go to the /usr/src folder and compile source code:

```
cd /usr/src
git clone https://stash.freeswitch.org/scm/fs/freeswitch.git
cd /usr/src/freeswitch
./bootstrap.sh -j
./configure --enable-core-pgsql-support
make && make install
```

 In this case, we will use the master version. Note that master versions are usually unstable, and for production systems, you should use stable versions only. For this information, refer to the home page and clone the relevant stable version at https://www.freeswitch.org.

3. Install sounds:

```
make cd-sounds-install cd-moh-install
```

4. Set permissions and the file owner:

```
cd /usr/local
adduser --disabled-password  --quiet --system --home /usr/local/
freeswitch --gecos "FreeSWITCH Voice Platform" --ingroup daemon
freeswitch
chown -R freeswitch:daemon /usr/local/freeswitch/
chmod -R ug=rwX,o= /usr/local/freeswitch/
chmod -R u=rwx,g=rx /usr/local/freeswitch/bin/*
```

 For more details, refer to https://www.freeswitch.org.

Enabling WebRTC

FreeSWITCH supports WebRTC from version 1.4. WebRTC can be enabled or disabled by changing appropriate options in the configuration of FreeSWITCH. By default, configuration options that enable WebRTC are commented out, so WebRTC is disabled. To enable WebRTC in FreeSWITCH, you should open sip_profiles/internal.xml configuration file and edit appropriate configuration options as shown:

```
<!-- uncomment for sip over websocket support -->
<param name="ws-binding"  value=":5066"/>

<!-- uncomment for sip over secure websocket support -->
<!-- You need wss.pem in /usr/local/freeswitch/certs for wss -->
<!--<param name="wss-binding" value=":7443"/>-->
```

You will need to restart FreeSWITCH after this change.

 You need to use SSL/TLS certificates if you want to
utilize the WebSockets secured layer (WSS).

Starting FreeSWITCH

You need to add a new user into FreeSWITCH. Please refer to the appropriate page on this
topic at `https://wiki.freeswitch.org/wiki/XML_User_Directory_Guide`.

After you've made all the configuration steps, start the FreeSWITCH by using the
following command:

```
cd /usr/local/freeswitch/bin
./freeswitch
```

Now we have FreeSWITCH installed with enabled with the support of WebRTC.

How it works...

It's better to use a diagram to describe the workflow, so have a look at the following diagram:

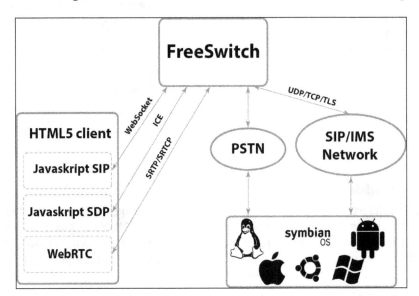

There's more...

FreeSWITCH is not the only VoIP platform solution existing in the world. One of the best-known
alternatives is Asterisk.

Deciding which particular solution might fit your requirements is all up to you. They both have support for WebRTC since the last versions (middle of 2014). So they both might contain some bugs or features related to the technology.

Asterisk seems to be older and more mature than FreeSWITCH. There are more hacks and there's more documentation related to Asterisk than FreeSWITCH.

So if you are looking for alternatives to FreeSWITCH, it might be worth trying Asterisk. Its home page is `http://www.asterisk.org`.

See also

▸ For an alternative solution, using other VoIP software, refer to the *Integrating WebRTC with Asterisk* recipe

▸ In the *Making calls from a web page* recipe, we will cover how to make calls from web pages using WebRTC and a VoIP platform integration

Making calls from a web page

In this recipe, we will cover the process of making calls from web pages. For this task, you will need to run a VoIP service. It can be your own Asterisk or FreeSWITCH installation, or it can be some external, cloud, or SaaS VoIP solution.

To achieve our goal, we will use an HTML5 SIP library to make calls from a web page to a phone number and vice versa.

Getting ready

In this recipe, we will work under Linux, so prepare a Linux box. We will also use tools such as Git and SVN—install them if they're not installed yet on your machine.

You will need a web server installed. It might be Nginx, Apache HTTP Server, or any other web server you like the most. I assume that you have some experience of installing and configuring Linux software.

How to do it...

During this recipe we will install and configure a set of applications and build a service by integrating these applications with each other. We will not cover all the installation and configuration steps from scratch, but will only cover specific steps that might be relevant to this recipe.

Installing sipML5

The first HTML5 SIP client is sipML5. We will use this library in this recipe to achieve our goal.

1. Go into your default www folder of the web server. It might vary on different systems. For Ubuntu it can be `/usr/local/www`.

2. Download the sipML5 source code:

   svn checkout http://sipml5.googlecode.com/svn/trunk/

3. Give Asterisk access rights to downloaded the project:

   chown -R asterisk:asterisk /usr/local/www/trunk/

4. Open the Chrome web browser and navigate to `http://<your_IP>/trunk/call.htm`.

   Here, `your_IP` is the IP address of your machine where sipML5 has been installed.

5. Go to **Expert Mode** and set the options as depicted in the following screenshot:

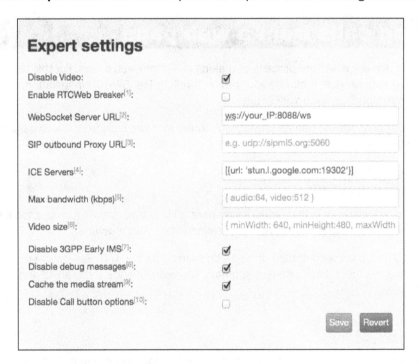

Put your actual machine's IP address instead of `your_IP`.

 If you have your own STUN server installed, you can specify its IP or name at the **ICE Servers** option.

6. Save the changes.

7. Now get back to the first tab and fill in the fields as depicted in the following screenshot:

Note that you should out your machine's actual IP address (where Asterisk and sipML5 are installed) instead of the your_IP word.

 Use the same password you configured for Asterisk (SuperS3cret in this recipe).

Now click on **Login**—you should see a **Connected** status line at the top of the **Registration** box.

Now you can try to make an outgoing call using the **Call** control—call on any number that is served by the VoIP platform (Asterisk or FreeSWITCH) and is registered in the system. Incoming calls should work as well; you can check them using any SIP softphone client. Here are a few of them:

Bria: For more information, go to http://www.counterpath.com/bria

Telephone: To know more about Telephone, refer to https://github.com/eofster/Telephone

Zoiper: For more details, refer to http://www.zoiper.com/en

Express Talk: Refer to `http://www.nch.com.au/talk/` for more information

3CXPhone: For more information, go to `http://www.3cx.com/voip/softphone/`

X-Lite: To know more about X-Lite, refer to `http://www.counterpath.com/x-lite`

How it works...

The working flow of this constructed software system might be looking relatively complex for someone who is not building such systems every day. Although, the working flow of the integrated system is not that complex:

▶ HTML5 SIP client (sipML5 in our case) is just a VoIP softphone implemented to run in the browser.

▶ The in-browser softphone uses WebRTC technology to get access to the computer's multimedia (camera and microphone).

▶ Then using WebRTC, SIP protocol, and WebSockets, the in-browser softphone establishes communication with the VoIP platform (Asterisk or FreeSWITCH for example). Then, the softphone registers in the system. After that, the softphone becomes available to the user to make calls.

▶ Thus, the in-browser softphone becomes able to make phone calls to other endpoints of the VoIP platform. If the VoIP platform has a gate to an external phone network, you can even make external phone calls using just the in-browser softphone.

There's more...

The sipML5 library is not the only solution that can be used for this task. There are several alternative software pieces that can be used in this scope as well. Here are two examples of them:

SIP.js: For more information, refer to `http://sipjs.com/`

JsSIP: Refer to `http://jssip.net/` for more information

Each library has its own pros and cons and can be suitable for your particular expectations and requirements. The common integration schema remains the same, so you can try different software and decide which one is best for you.

See also

You will need a VoIP platform (SIP server) installed to make calls from a web page. You can use an existing external server or you can install your own. To install your own VoIP platform, please refer to the following recipes:

- Refer to the *Integrating WebRTC with Asterisk* recipe to learn how to integrate WebRTC with Asterisk

- Refer to the *Integrating WebRTC with FreeSWITCH* recipe to learn how to integrate WebRTC with different VoIP solutions such as FreeSWITCH

 It would probably be good idea to use an external or cloud VoIP platform for such purposes in production. Maintaining a good, working, and scalable VoIP platform cannot be easy.

Integration of WebRTC with web cameras

In this recipe, we will discuss how to integrate WebRTC with web cameras. Why might someone want to integrate a web camera with WebRTC technology? Here are some reasons why they might do this:

- A web camera needs a Java or ActiveX enabled on the client for it to be able see the image from the camera. Many computers have Java installed; nevertheless in some cases, it might be impossible to install/use Java or ActiveX. Regarding ActiveX, this technology is supported even on fewer devices than Java. WebRTC can become a universal and lightweight way to show multimedia from a webcam and that doesn't need you to install any additional software.

- As of now, WebRTC is fully supported on Android devices (mostly the ones that use Chrome mobile), but in the near future, it is supposed to be supported on other mobile platforms as well (such as iOS and Windows Mobile). At this time, you usually have to install JVM or FlashPlayer in your mobile if you want to see a video from a webcam. Often, it is barely possible at all.

- Webcams usually are very resource limited devices. When several clients access the camera at one time, it can show time delays and can even get stuck. Such an issue can be solved very effectively by using of a WebRTC application that is integrated into the connection between the user and the camera.

Here we cover possible solution for such a task: capturing a video from a webcam, transcoding it into WebRTC flow, and displaying it in the web browser.

Getting ready

There are many ways in which webcams give out videos. Usually, it can be a set of JPEG images or RTSP flow. In our experiments, we will cover the second case and will use a D-Link DCS-5220 web camera.

So for this recipe, you need a webcam that can do RTSP. In my case, it is D-Link but you can use any other webcam— the recipe will still be relevant, but some minor changes might be necessary. Install and configure the webcam and connect it to the network.

In this recipe, we will also install and configure the WebRTC media server—this software is written in Java, so you need JVM installed in your box. One more thing that you will need to do is install a web server. You can use Nginx, Apache HTTP Server, or any other web server of your choice.

How to do it...

We will configure the webcam. Then we will install and configure the WebRTC media server, and then we will connect all the components in the whole system.

Configuring the webcam

First of all, we will do some minor configurations with the web camera. To do so, perform the following steps:

1. Navigate to the webcam's admin page and open the **NETWORK SETUP** menu. We need to go to the RTSP section:

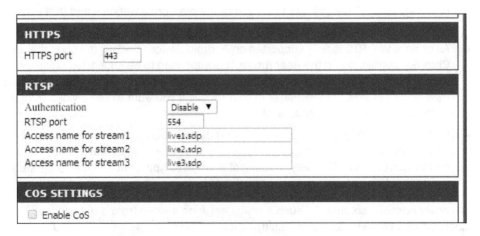

In this section, we need to look for the **RTSP port** parameter—it should be **554** by default. It is also worth to set the RTSP **Authentication** field to the **Disable** state—for the time being we're working on the task.

Check whether the webcam works as expected. For this, you can use VLC media player—just open `rtsp://cam_IP/live1.sdp` in the player.

> Note that you need to insert the relevant IP address of the web camera instead of `cam_IP`. If the camera is configured the correct way, you will see a video captured from it.

Installing WebRTC media server

As we know already, our web camera streams media over RTSP, but we want to watch that stream in a web browser using WebRTC. So you have to convert the media from RTSP to the WebRTC form. For this purpose, we will use the WebRTC media server from Flashphoner.

This software can capture media from RTSP streamer, re-encode it, and stream it in WebRTC:

1. Download the media server from its home page at `http://flashphoner.com/download_webrtcserver/`.

2. Unpack the archive:

   ```
   tar -xvzf FlashphonerMediaServerWebRTC.tar.gz
   ```

3. Install the server:

   ```
   cd FlashphonerMediaServerWebRTC
   ./install.sh
   ```

> During the installation, you will be asked on the public and private servers' IPs. If you're experimenting on your local machine, both the IPs might be identical.

4. Start the media server:

   ```
   service webcallserver start
   ```

5. Check whether that server is running:

   ```
   ps -ax | grep Flashphoner
   ```

> You also can look into your media server's log files to check whether everything is all right: `/usr/local/FlashphonerWebCallServer/logs/server_logs/flashphoner.log`.

6. Go to your web server's www folder—in my case it is `/usr/local/www`:

   ```
   cd /usr/local/www
   ```

7. Download the web UI files into the folder:

```
wget https://github.com/flashphoner/flashphoner_client/archive/
wcs_media_client.zip
```

   Clients will access this UI via the web server in order to see the captured media streams from the web camera. In other words, this is the UI for the media server.

8. Unpack the archive:

```
unzip wcs_media_client.zip
```

9. There are several nested empty folders in the archive, so it is worthwhile moving the necessary files to the upper level and making life a bit easier with the following commands:

```
mv flashphoner_client-wcs_media_client/client/wcs_media_client ./

rm -rf flashphoner_client-wcs_media_client/
```

10. Edit this `wcs_media_client/flashphoner.xml` configuration file and set the proper IP address of the WebRTC media server:

```
<flashphoner>
    <wcs_server>188.226.144.63</wcs_server>
    <ws_port>8080</ws_port>
    <video_width>1280</video_width>
    <video_height>720</video_height>
</flashphoner>
```

The media server is now installed and properly configured!

Time for magic

Now when everything is configured and running, it is time to do the magic. From your web browser, go to `http://<server_IP>/wcs_media_client/?id=rtsp://<cam_IP>/live1.sdp`.

The following parameters are mentioned in the preceding URL:

- `<server_IP>`: This is the IP address of the machine where the WebRTC media server with its UI is installed
- `<cam_IP>`: This is the IP address of the web camera

While navigating to the URL, you will first see an image from the media server, as shown in the following screenshot:

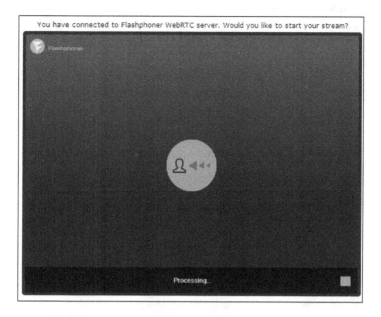

At this stage, the WebRTC media server will try to connect to the camera and negotiate with it regarding the stream capturing. It can take several seconds. When the communication process is done, the server begins capturing the media stream from the camera and encoding it into the WebRTC format. After that, you will see the image from the camera.

How it works...

The following diagram depicts the general schema of what we built in this solution:

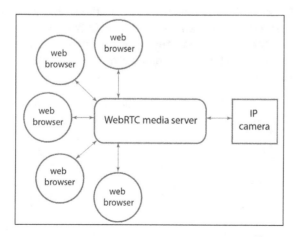

As you can see, the WebRTC media server captures the stream from the web camera and then the clients can see the captured stream in their web browsers using WebRTC. What is important here, is that clients are not connected to the webcam and they don't get media streamed from the webcam directly; instead, clients are connected to the WebRTC media server, and they get all media streams from the media server.

In the following diagram, you can see the workflow of how it works, step by step:

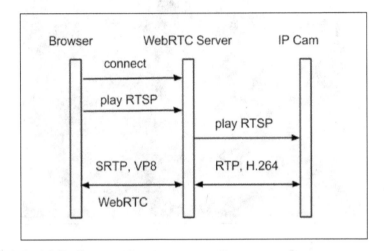

There's more...

You might want to take a look at another solution—janus-gateway. For more information refer to `https://github.com/meetecho/janus-gateway`.

This solution is open source (while the server from Flashphoner is not). At the time of writing this, it works under Linux only, but its authors claim cross-platform support in the future.

Another popular media server, Wowza, can also capture the RTSP stream from cameras, but its main purpose to re-encode media data into Flash, so for WebRTC, this solution is hardly suitable. Nevertheless, Wowza can be an interesting solution as well, for example, if you need your application to support Flash technology along with WebRTC. This software can be found at `http://www.wowza.com`.

Many cameras stream to Motion JPEG, and this recipe is irrelevant for such devices. Nevertheless, it is possible to build a similar solution for them as well, using similar schema.

4
Debugging a WebRTC Application

In this chapter, we will cover the following topics:

- ▸ Working with a WebRTC statistics API
- ▸ Debugging with Chrome
- ▸ Debugging TURN
- ▸ Debugging using Wireshark

Introduction

Debugging is a very important aspect in developing a computer software. Even if you are an experienced developer and write very clean and professional code, you might face some situations when the only good way to understand what's going wrong is debugging and profiling.

In this chapter, we will cover debugging within the scope of developing WebRTC applications. We will talk about specific useful tools built in Chrome web browser, which can be helpful. Also, we will cover basic questions of debugging JavaScript applications in the scope of the main topic. Of course, we will cover the server side as well.

WebRTC has a very useful API known as **statistics API**; it can be used for monitoring and debugging WebRTC applications. We will cover this topic in the appropriate recipe, considering real-world use cases and practical possible solutions.

A WebRTC application usually works very intensively with network. Therefore, we will learn how to use Wireshark (a network sniffer) for debugging purposes in the scope of developing WebRTC applications and services.

Working with a WebRTC statistics API

WebRTC's standard describes statistics API—a mechanism that an application can use for getting many kinds of statistical data. Using this mechanism can be helpful when debugging applications, because you can get access to some hidden data that is not visible to the application or to a customer in any other way.

Using this part of API you can better understand what is going on under the hood of the web browser and your application. It is very useful if you are a beginner and would like to know more on how all this works. It is also helpful if you're an experienced developer and are creating some advanced feature in the application.

Getting ready

For this recipe, we will not do much configuration work. We will not install any libraries or compile Linux software like we do in some other recipes. This recipe is dedicated to debugging and most of the topic is dedicated to client side. Therefore, most of the material is about JavaScript, the web browser and browser's console.

I would recommend you use Chrome for this recipe, because this browser still seems to be more stable in the scope of supporting WebRTC. Moreover, usually Chrome has better and more advanced support for this technology.

How to do it...

For accessing the statistics data, you should use the `getStats` API function (a method of `PeerConnection` instances). While calling this function, you have to pass the selector. In reply, the browser will return relevant statistical data.

Since WebRTC is still under development, the API functions might still have different names in the supported web browsers. To solve this issue, it is worthwhile to write additional code that could serve as a wrapper and universal API to the function. The following code can be used as a simple example of such behavior:

```
function myGetStats(peer, callback) {
    if (!!navigator.mozGetUserMedia) {
        peer.getStats(
            function (res) {
                var items = [];
                res.forEach(function (result) {
```

```
                    items.push(result);
                });
                callback(items);
            },
            callback
        );
    } else {
        peer.getStats(function (res) {
            var items = [];
            res.result().forEach(function (result) {
                var item = {};
                result.names().forEach(function (name) {
                    item[name] = result.stat(name);
                });
                item.id = result.id;
                item.type = result.type;
                item.timestamp = result.timestamp;
                items.push(item);
                items.push(item);
            });
            callback(items);
        });
    }
};
```

Now let's write a function that we will call from the application to get the statistics. This function will print statistical data to the browser's console every 5 seconds:

```
function printStats(peer) {
    myGetStats(peer, function (results) {
        for (var i = 0; i < results.length; ++i) {
            console.log(results[i]);
        }
        setTimeout(function () {
            printStats(peer);
        }, 5000);
    });
}
```

Next, we should put the function call in the proper place in the application. Somewhere in your application, you should create a peer connection object using a construction similar to the following:

```
pc = new RTCPeerConnection(pc_config, pc_constraints);
```

After that, you should set up the `onaddstream` callback of the created object:

```
pc.onaddstream = onRemoteStreamAdded;
```

Here, `onRemoteStreamAdded` is a callback function that is called once when peer connection is established. In the following callback function, you should add some code that calls the `printStats` function, which we have just written in the preceding code:

```
var onRemoteStreamAdded = function(event) {
        clog("Remote stream added.");
        attachMediaStream(remoteVideo, event.stream);
        remoteStream = event.stream;
        printStats(pc);
};
```

I have provided the full list of the functions here to show the big picture and make it clear. You can see in the following screenshot that after the media stream is attached to the proper video HTML tag, we call `printStats` so that it prints the statistical data to the console every 5 seconds:

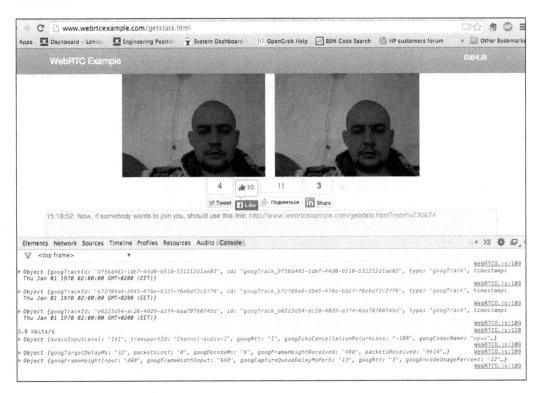

Here you can see a screenshot of an example web page that uses the described `printStats` function. The web browser console is opened, and you can see the statistical data printed there. The statistical data looks incomprehensible, but the following screenshots will give you more details, making it clearer.

The following screenshot depicts a part of the browser's console with one of the expanded statistic data objects. In the screenshot, you can see the `Object` structure, and according to its options it is an audio track: its input level is `131`, its used codec is Opus, and there were around 300 kilobytes sent through this channel. You can also see other useful information regarding this object, such as echo cancellation feature details.

```
Q  Elements  Network  Sources  Timeline  Profiles  Resources  Audits  Console
⊘  ▽  <top frame>                    ▼
  ▼ Object {audioInputLevel: "131", transportId: "Channel-audio-1"
      audioInputLevel: "131"
      bytesSent: "306542"
      googCodecName: "opus"
      googEchoCancellationEchoDelayMedian: "-60"
      googEchoCancellationEchoDelayStdDev: "0"
      googEchoCancellationQualityMin: "-1"
      googEchoCancellationReturnLoss: "-100"
      googEchoCancellationReturnLossEnhancement: "-100"
      googJitterReceived: "0"
      googRtt: "4"
      googTrackId: "9f56a461-1db7-44d0-b518-531212d1ae83"
      googTypingNoiseState: "false"
      id: "ssrc_2532551919"
      packetsSent: "3987"
      ssrc: "2532551919"
```

Another screenshot presents one more expanded statistic object. In the following screenshot, you can see that we deal with video data, we have a delay of 33 milliseconds, and the frame size is 640 x 480. More service information is present in the following screenshot:

```
Q  Elements  Network  Sources  Timeline  Profiles  Resources  Audits  Console
⊘  ▽  <top frame>                    ▼
  ▼ Object {googTargetDelayMs: "33", packetsLost: "0", googDecodeMs: "6"
      bytesReceived: "11683103"
      googCurrentDelayMs: "33"
      googDecodeMs: "6"
      googFirsSent: "0"
      googFrameHeightReceived: "480"
      googFrameRateDecoded: "0"
      googFrameRateOutput: "0"
      googFrameRateReceived: "0"
      googFrameWidthReceived: "640"
      googJitterBufferMs: "15"
      googMaxDecodeMs: "8"
      googMinPlayoutDelayMs: "0"
      googNacksSent: "-1"
      googRenderDelayMs: "10"
      googTargetDelayMs: "33"
```

Let's see one more example screenshot. In the following screenshot, we can see that the used video codec is VP8, the video frame size is 640 x 480, and around 13 megabytes of video data have been sent through this media channel:

The `getStats` WebRTC API function can be very useful not only for debugging purposes. This function can be helpful for many use cases, for example:

► **Monitoring**: In this use case, if you have your web service running, you probably want to monitor its state dynamically, to know how well the resources are utilized and so on

► **Tests**: For this use case, if you're working on some feature or just implementing some new functionality in your application, statistics API can be helpful with A/B testing

► **Troubleshooting**: In this use case, if your application doesn't work by some reason for a customer, you can use this mechanism to track the issue and find the root cause

Checking estimated bandwidth

We just considered a common case of using WebRTC statistics API. Now we will consider a practical example of using this mechanism. In particular, we will try to know our estimated bandwidth for the video channel used in our application.

The following function collects statistical data related to the bandwidth utilization and prints a simple report on the console:

```
function printStats(peer) {
```

The `myGetStats` function is described as follows and can be found in the *How to do it...* section of this recipe:

```
myGetStats(peer, function (results) {
    for (var i = 0; i < results.length; ++i) {
        var res = results[i];
```

Check if we have a video object:

```
        if (res.googCodecName == 'VP8') {
            if (!window.prevBytesSent) window.prevBytesSent =
            res.bytesSent;
```

Get the `bytesSent` value as follows:

```
            var bytes = res.bytesSent - window.prevBytesSent;
            window.prevBytesSent = res.bytesSent;
```

Now convert the value into kilobytes:

```
            var kilobytes = bytes / 1024;
            console.log(kilobytes.toFixed(1) + ' kilobytes per
            second');
        }
    }
    setTimeout(function () {
        printStats(peer);
    }, 1000);
});
```

We have set the timeout value to 1,000 milliseconds. Thus every second this function gets statistics using WebRTC API, extracts the sent bytes value from the appropriate object, and calculates the bitrate. The following screenshot depicts what you should see in the browser's console:

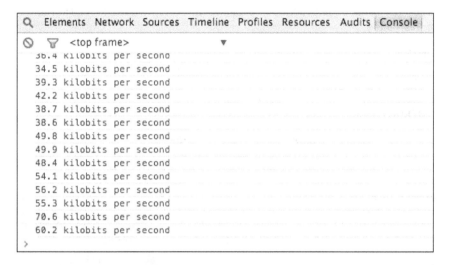

The following section represents one more use case that you might face while developing an application or a service using WebRTC features.

Checking packet loss

In this section, we will consider another use case: checking packet loss. This is an example taken from the WebRTC standard draft, a bit adapted to our code base. In the scenario, the user is experiencing bad sound, and the application wants to determine whether packet loss causes this issue with the following steps:

1. First of all, let's declare the variables where we will store baseline values and current value:

   ```
   var baselineReport, currentReport;
   ```

2. Next, write initialization function—it will make first call to statistics API and store the first value as baseline:

   ```
   function initStats (peer) {
       myGetStats(peer, function (report) {
           baselineReport = report;
       } );
   ```

3. Now, using timer, we will get statistics every one second and process it:

   ```
   setTimeout(function () {
       myGetStats(peer, function (report) {
   ```

```
            currentReport = report;
            processStats();
        });
    }, 1000); }
```

4. The following function does all the processing work:

```
function processStats() {
    // compare the elements from the current report with
    the baseline
    for each (var now in currentReport) {
        if (now.type != "outbund-rtp") continue;
        // get the corresponding stats from the baseline
        report
        base = baselineReport[now.id];
        if (base) {
            remoteNow = currentReport[now.remoteId];
            remoteBase = baselineReport[base.remoteId];
            var packetsSent = now.packetsSent -
            base.packetsSent;
            var packetsReceived = remoteNow.packetsReceived
            - remoteBase.packetsReceived;
            // if fractionLost is > 0.3, we have probably
            found the culprit
            var fractionLost = (packetsSent -
            packetsReceived) / packetsSent;
            if (fractionLost > 0.3) {
                console.log("fractionLost is too big: " +
                fractionLost); }
        }
    }
}
```

5. Now, the following code represents how all that we just have written can be used in the application:

```
var onRemoteStreamAdded = function(event) {
    clog("Remote stream added.");
    attachMediaStream(remoteVideo, event.stream);
    remoteStream = event.stream;
    initStats(pc);
};
```

Here, we will call the `iniStats` function. This function will get the first data from the statistics API; store it in the memory, and set up a time for one second. Then, every second another function will be called—it will get the next statistics sample and do calculations trying to determine if something is wrong with the packet loss value.

How it works...

The web browser collects and maintains a set of statistic data that can be accessed via WebRTC API. When accessing this data, you should use a selector—something that determines the kind of data you want to retrieve.

The selector might, for example, be a `MediaStreamTrack` object. In this case, the valid selector must be a member of a `MediaStream` object that is sent or received by the `PeerConnection` object, for which statistics is requested.

 Using the selector and calling the `getStats` function, you will get statistics data packed in a JavaScript object. Then you need to parse it and get the necessary value. Most WebRTC API functions allow you to set up an error function callback. This function will be called if something goes wrong; usually such callback functions serve to print error messages in a console. Using these error callbacks is mandatory. Even if you don't pass the error callback and everything works well, the situation might change with the next browser update, and your application will throw an exception. Therefore don't miss the error callbacks!

There's more...

For more details, refer to WebRTC standard draft at `http://dev.w3.org/2011/webrtc/editor/webrtc.html`, where you can find more information regarding this part of API. The standard is in the draft stage yet, so some (or many) concepts might be changed.

See also

▸ Take a look at the *Debugging with Chrome* recipe. Chrome has a set of built-in WebRTC-related tools that might be helpful when developing and debugging WebRTC applications.

Debugging with Chrome

Chrome is a web browser developed by Google—the company that invests in WebRTC development very intensively. Chrome usually has the most advanced support of WebRTC features than other browsers, and new and experimental features usually appear first in Chrome.

Thus, it is not surprising that Chrome has good tools for debugging the WebRTC stack. Some of the relevant details will be covered in this recipe.

Getting ready

For this recipe, you will need Chrome installed. It is a multiplatform, so you can download the relevant installation pack from its home page at `https://www.google.com/chrome/browser/`.

How to do it...

There are two known Chrome mechanisms that can be useful for debugging WebRTC applications:

- WebRTC-internals
- Logging

In most cases, you probably will use the first one.

Using webrtc-internals

WebRTC-internals is a built-in mechanism in Chrome with the use of which you can get access to a variety of WebRTC stack-related information and statistics data.

Open a Chrome web browser and go to the URL `chrome://webrtc-internals/`.

If you haven't opened any WebRTC application yet, you will not see anything interesting. Now in the new tab, open a web page of a web application where a WebRTC API is utilized, and refresh the page that has opened *web-internals*. You will see something similar to what is depicted in the following screenshot:

Here you can see the screenshot of a real application; its URL is present at the top of the window. In the brackets, you can see the list of STUN/TURN servers that the web browser uses for establishing peer-to-peer connection. There, also shown are the optional parameters that are specified while creating that peer connection, for example, the **DtlsSrtpKeyAgreement** option.

Below the list there are several lines with horizontal arrows that can be expanded, and there you will find additional details regarding the application and WebRTC stack. There is not much information that can be displayed because at this stage the direct peer-to-peer connection is not established yet.

The following screenshot depicts the next stage right after establishing the peer-to-peer connection:

Here you can see more lines; each of them represents data related to some object or event. The following screenshot shows an example of what kind of data you can find while expanding these lines:

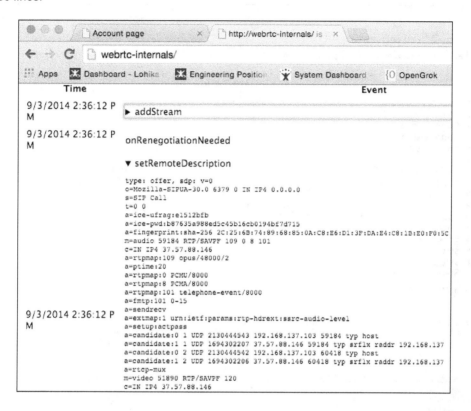

You can see that I've expanded the `setRemoteDescription` list item, and there are details that have appeared for this object: this is an SDP message of the type `offer`. You can also see relevant information about the candidates, codecs, and IP addresses of this item.

In the next screenshot, you can find even more examples of different kinds of items that can be accessible via this page:

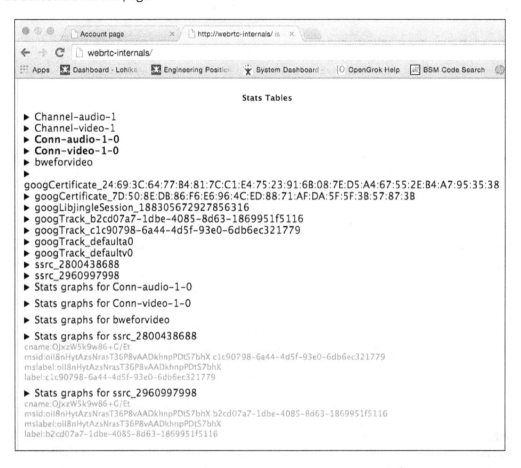

Now here are the audio and video connection objects available and many other service items that are not obvious. Let's see what is under the audio connection item in the following screenshot:

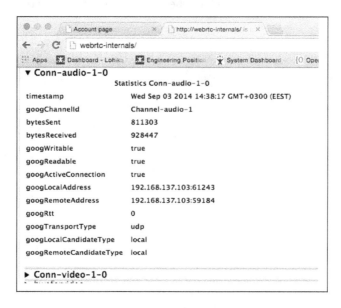

Here we expanded the **Conn-audio** object that represents the audio connection. You can see the bytes that were sent and received, IP addresses of peers (I was running this example on my notebook locally, so both IP addresses are identical), transport protocol type, and other options.

You will see the same kind of information while expanding the video connection item, so I will skip the screenshot for this one. Instead of that, let's see what is on **bweforvideo**:

This item represents the bandwidth-related details. Here you can find the bitrate and bandwidth utilized by the web browser during the communication.

In the following screenshot, you can find another example related to video data:

What is good with this tool is that it gives not only numbers and raw data, but it also presents great-looking graphics, where you can visually see what is happening. In the following screenshot, you can see the graphs related to audio and video channels utilization:

Now let's take a look at another graphic representation—**bweforvideo**. It represents various network connection parameters related to the video channel. On the left-hand side, you can find options through which you can enable or disable the parameters that you want or don't want to see in the graphic representation.

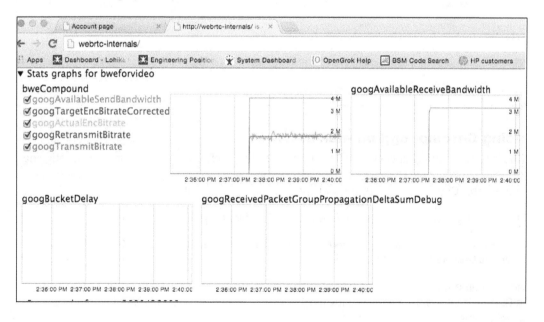

There are more graphic representations available—every graphic represents a dynamical change in some parameter.

Using Chrome logging mechanism

This is not something specific to WebRTC, but can be helpful while developing and debugging WebRTC applications. Chrome can be started with enabling the logging for certain modules. In this case, Chrome during its work will print a variety of useful details into log files.

The following command starts Chrome with enabled logging:

```
chrome --enable-logging --v=4 --vmodule=*libjingle/source/talk/*=4
--vmodule=*media/audio/*=4
```

Now, Chrome will put additional details into the `chrome_debug.log` file that can be found in Chrome's user data folder. The log file is a plain text file, so you can read it without using special tools.

On some systems, this log file might be directly written into the terminal.

Although we are working on the log file under Windows, you can use convenient tools such as Sawbuck. You can find its home page at `https://code.google.com/p/sawbuck/`.

Sawbuck is a log files viewer that can be used not only for Chrome logs, but also for working with logs of other applications (using plugins). You can see what this tool looks like in the following screenshot (taken from the tool's home page):

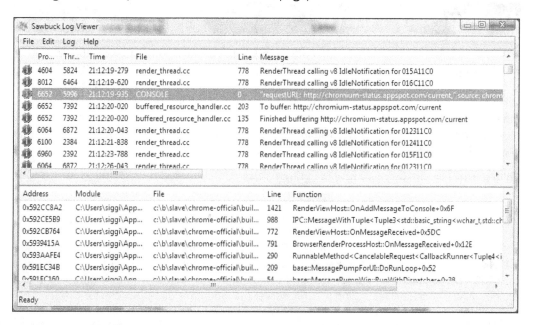

How it works...

We learned the built-in mechanism available in Chrome that can help debugging and profiling while developing WebRTC applications. Chrome collects useful data, and using logging and the webrtc-internals tool, you can access these data. Moreover, by accessing graphs, you can analyze the process in a dynamic manner.

To use this tool, you don't need to install any additional software. This makes it irreplaceable in the application development process.

There's more...

You can find more details specific to Chrome by logging on the appropriate web page of the Chromium project at `http://www.chromium.org/for-testers/enable-logging`.

See also

▶ For server-side debugging advices, please refer to the *Debugging TURN* recipe

Debugging TURN

As you probably know, your application will definitely use STUN if you want it to work in the real world. Using STUN will be enough for most cases, although you will have to use TURN in many situations—especially when working with enterprise customers, because they usually have very strict network firewall policies and complex network configurations. Using TURN can be the only available solution for customers located in some places, for example, some countries might have specific network access limitations that cause issues for network applications that are WebRTC-based.

So in this recipe, we will cover how to debug TURN.

Getting ready

For this recipe, you need to have your own TURN server installed and running. When you use a TURN server as a third-party service, you can debug only client side. However, if you use your own TURN server, you have access to it and can do more in the scope of debugging. So in this recipe, we will consider debugging a TURN server that you have direct access to.

How to do it...

In *Chapter 3*, *Integrating WebRTC*, we considered the installation and configuration of our own TURN server. To debug TURN, set the verbosity level to maximum and run the TURN server in console. Then start your WebRTC application using the TURN server—when the application will contact the server, you will see debug messages on the console display where the server is running. The following represents the kinds of messages you might see in the console:

```
129: session 128000000000000001: new, username=<user1:alpha>,
lifetime=3600

129: session 128000000000000001: user <user1:alpha>: incoming packet
ALLOCATE processed, success

129: handle_udp_packet: New UDP endpoint: local addr 176.58.121.75:3478,
remote addr 89.209.127.164:50186
```

```
130: session 128000000000000007: user <>: incoming packet BINDING
processed, success

130: session 128000000000000009: user <>: incoming packet message
processed, error 401

131: session 128000000000000009: new, username=<user2:beta>, lifetime=600

131: session 128000000000000009: user <user2:beta>: incoming packet
ALLOCATE processed, success

131: handle_udp_packet: New UDP endpoint: local addr 176.58.121.75:3478,
remote addr 89.209.127.164:52914

131: session 128000000000000010: user <>: incoming packet message
processed, error 401
```

In this dump, you will see a fragment of TURN authentication stage where two clients are trying to get authenticated. Session 129 represents the client `user1` with the `alpha` password, and session 131 represents the customer `user2` with the `beta` password. You can also see session 130, which represents a STUN client—it doesn't use TURN functionality, so you don't see any usernames or passwords from this client.

Now if you've configured the TURN server with default console options, you can connect to the TURN console and get more specific details on the certain session. Connect to the TURN console:

```
telnet localhost 5766
```

After you've connected, it will show you something like the following:

```
Connected to localhost.

Escape character is '^]'.

TURN Server

rfc5766-turn-server

Citrix-3.2.2.910 'Marshal West'

Type '?' for help
```

In the console you have a set of commands—using ? or `help` you can ask the system to show the whole list of available commands and options. The command we're interested in is `ps`—it shows detailed information about the available TURN/STUN sessions.

```
> ps


  7) id=128000000000000004, user <user1:alpha>:
     started 78 secs ago
     expiring in 3522 secs
     client protocol UDP, relay protocol UDP
     client addr x.x.x.x:58454, server addr y.y.y.y:3478
```

```
        relay addr x.x.x.x:63599
        fingerprints enforced: ON
        mobile: OFF
        SHA256: OFF
        SHA type: SHA1
        usage: rp=2, rb=172, sp=1, sb=120
         rate: r=0, s=0, total=0 (bytes per sec)


    8) id=128000000000000010, user <user2:beta>:
        started 76 secs ago
        expiring in 524 secs
        client protocol UDP, relay protocol UDP
        client addr x.x.x.x:52914, server addr y.y.y.y:3478
        relay addr x.x.x.x:50796
        fingerprints enforced: OFF
        mobile: OFF
        SHA256: OFF
        SHA type: SHA1
        usage: rp=2, rb=140, sp=1, sb=120
        rate: r=0, s=0, total=0 (bytes per sec)


  Total sessions: 8
```

From this listing we can see that in total there are eight sessions on the server. In this preceding fragment, we see details on certain two sessions. We know the usernames (`user1` and `user2`), passwords, IP addresses, time of expiration, time of living, and some more details of each session.

Using the TURN console, you can check whether some problematic client has connected to the server successfully or has any issues. You can check which usernames or passwords have been used for each session. You can also know about the used protocols and encryption details. Analyzing such kinds of information can help in troubleshooting the TURN/STUN communication process.

How it works...

Having direct access to the TURN server, you can use its console to get more certain data and analyze what's going on. Using such a method, you can debug your application that is using TURN.

There's more...

In this recipe, we considered a certain way to implement a TURN server, using rfc5766-turn-server software. If you use some other software, it might be supplied with some other specific tools for debugging and diagnostic.

See also

► When you have no direct access to the TURN server, you can use a network sniffer to capture network packets and analyze the situation from that side. To learn this technique, please refer to the *Debugging using Wireshark* recipe.

► To configure and install a TURN server, refer to *Chapter 3, Integrating WebRTC*.

Debugging using Wireshark

WebRTC applications use networks very intensively. Thus sometimes you might need to debug not just the application, but also its communication with other components of the whole system.

In this section, we will cover the process of debugging WebRTC applications using network sniffer.

Network sniffer is a tool for capturing network packets. Usually, such tools can help you to analyze captured data. Using sniffer, you can see and understand how your application communicates with other points.

Getting ready

For our recipe, we will use Wireshark—which is a free and multiplatform network sniffer software. Download it from the home page at `http://www.wireshark.org`.

This tool is very user-friendly and works on most popular platforms, so you don't need any specific preparations.

You will also need some WebRTC application; you can use any simple hello world application for this purpose.

How to do it...

Start Wireshark. You will see a UI that might look confusing at the first time. This tool is very powerful and has many features, but for this task, we will use basic functionality. Perform the following steps to use Wireshark:

1. Click on the **Capture** button—Wireshark will begin capturing data network frames.

2. Start your WebRTC application and navigate Chrome browser to the application's main web page.

3. Navigate to the application's web page using another browser and make a call to the first peer.

4. Wait until the WebRTC session begins and click on the **Stop** button in the Wireshark's UI.

Now let's see what we can get from the collected data. In the following screenshots, you can see the examples from my machine.

In the first screenshot, you can see a set of network packets that are sent between peers (my notebook and another work machine). The selected line points to a STUN binding success response with following decoded fields:

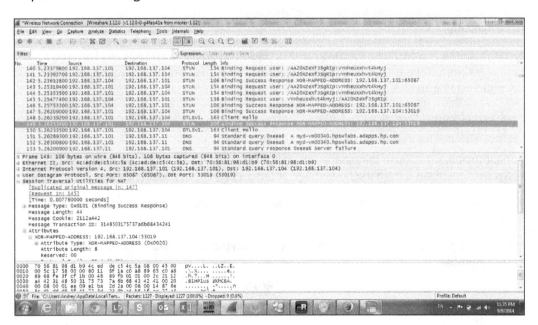

After that, peers try to establish secure direct connection, and you can see this stage in the following screenshot:

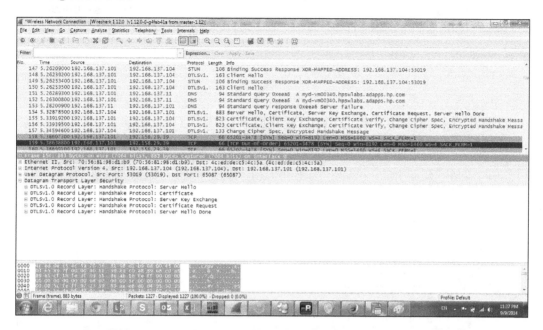

Another example of communication through secured channel is depicted in the following screenshot. Here you can see the application's data exchanging stage.

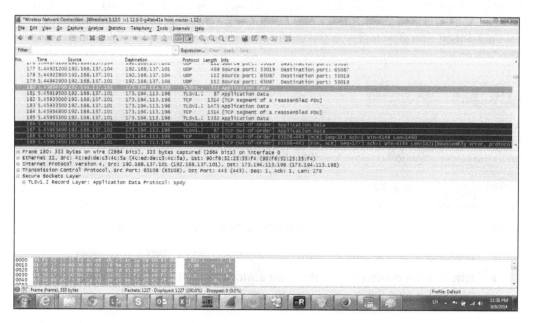

The following screenshot depicts the TURN authentication stage. You can see that the server replied with 401 unauthorized request; this is normal step at this stage and it just means that the server will not serve for anonymous client. After getting this server's response, the client will continue the communication process and will send credentials to the server.

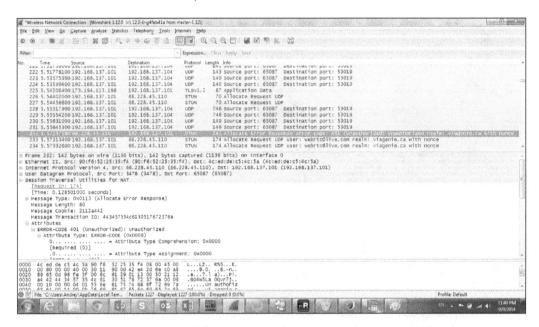

Using a network sniffer, such as Wireshark, can be very useful and helpful in the debugging process. You need to capture network packets during a certain stage of the application's communication, and after that, you can analyze the communication process to understand what's wrong.

How it works...

A network sniffer allows you to capture necessary network packets that are being sent between peers and servers. By analyzing these packets, we can understand what's going on in the communication channels and fix the issues.

There's more...

There are other network-related tools that might be helpful for such kind of task:

- **tcpdump**: This is a console network sniffer standard for UNIX-like systems
- **mtr**: This is a network tool that can be useful when you need to analyze a network path of the data that is sent between peers

See also

▶ When debugging network-related issues, using the webrtc-internals mechanism might also be useful. Refer to the *Debugging with Chrome* recipe for the details.

▶ Regarding the process of debugging TURN servers, you can refer to the *Debugging TURN* recipe.

5
Working with Filters

In this chapter, we will cover the following topics:

- ▶ Working with colors and grayscale
- ▶ Working with brightness
- ▶ Working with contrast
- ▶ Working with saturation
- ▶ Working with hue
- ▶ Using the sepia filter
- ▶ Using the opacity filter
- ▶ Inverting colors
- ▶ Implementing the blur effect
- ▶ Implementing the dropped shadow effect
- ▶ Combining filters
- ▶ Custom video processing

Introduction

With the introduction of the HTML5 standard, we have got new powerful features. One of the interesting ones is a CSS filter. Using this feature, you can control a variety of an image's properties. You can process a static image or video image on the fly.

In the scope of WebRTC, usage of filters enables you to implement new features in your application; it can control video images, make it brighter or less contrast, and apply some specific kinds of filters.

In this section, we will cover using of image processing, implementing several practical solutions and utilizing video filters. You will see *before* and *after* cases presented in the screenshots.

[🔅 This feature is not supported by all web browsers—use Chrome browser while testing the provided examples.]

The work on HTML5 and WebRTC standards is not finished yet, so there is a chance that certain places in the code might need to be changed in future. Note that these filters can only be applied locally. This means that during a video conference, if you apply a filter to the video from your web camera, you will see the changes locally in your browser—but your peer won't see these changes. It will see the original video translated from your web camera. On the other side, you can apply these filters to the remote video of your peer that is shown in your web browser.

You can find the source codes of the demo application supplied with this book.

Working with colors and grayscale

This recipe shows how to work with a filter that deals with the colors of the processed video. We will make a video less colorized and then make it black and white. This recipe can be used as a kind of simple special effect for a video.

How to do it...

Perform the following steps:

1. Add the control button to the main web page of your application:

   ```
   <button onclick="doGrayScale()">do grayscale</button>
   ```

2. Add an appropriate JavaScript function:

   ```
   function doGrayScale() {
     var v = document.getElementById("localVideo");
     v.style.webkitFilter="grayscale(50%)";
   };
   ```

   Here, `localVideo` is the ID property of the HTML video tag for the local video playback.

3. Navigate your browser to the web page. You will first see an unprocessed video from the web camera. The following screenshot depicts such a situation:

4. Now click on the **do grayscale** button—you will see that the image has become less colorized, as shown in the following screenshot:

This happened because we applied the grayscale filter with a value of 50%. In other words, we removed 50 percent of colors from the video.

5. Now edit the code and put 100% into the filter's value, reload the web page, and click on the **do grayscale** button again—you will see that video becomes black and white.

How it works...

When you click on the **do grayscale** button, the JavaScript function from the second step of the *How to do it...* section is called. This function applies the grayscale filter with the appropriate value to the video object—using its style HTML property. From now on, the web browser will show this video applying the filter on the fly.

Working with brightness

This recipe shows how to change the brightness of a video using the HTML5 filter. If you develop a video application, it's usually a good idea to give some control on the video to customers, allowing them to change the contrast, brightness, and other parameters of the video.

How to do it...

Follow the given steps:

1. Add the following control element to the main web page of your application—using this object we will change the brightness:

```
Brightness
<input type="range"
oninput="changeBrightness(this.valueAsNumber);
" value="0" step="0.1" min="0" max="10">
```

2. Add the appropriate JavaScript function:

```
function changeBrightness(val) {
  var v = document.getElementById("localVideo");
  v.style.webkitFilter="brightness(" + val + ")";
};
```

3. Here, localVideo is the ID property of the HTML video tag for the local video playback.Navigate your web browser to the web page. You will first see an unprocessed video from the web camera. The following screenshot depicts such a situation:

4. On the left-hand side of the page, you can see a control described as **Brightness**. Try to move it a little to the right—you will see that the video is becoming brighter. In the following screenshot I moved the control too much to the right and the image became too bright:

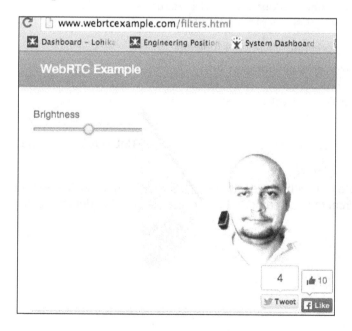

5. If you move it too much to the left, you will just see a black box.

How it works...

When you move the control, the JavaScript function from the second step is called. This function applies the `brightness` filter with the control's value to the video object—using its style HTML property. From now on, the web browser will show this video applying the filter on the fly.

Working with contrast

This recipe shows how to control the contrast feature of a video using the HTML5 filter feature. This is the second most important control that customers usually want to have when using video applications.

How to do it...

Follow the given steps:

1. Add a control element to the main web page of your application—using this object we will change the contrast:

   ```
   Contrast
   <input type="range"
   oninput="changeContrast(this.valueAsNumber);"
   value="0" step="0.1" min="0" max="10">
   ```

2. Add an appropriate JavaScript function:

   ```
   function changeContrast(val) {
     var v = document.getElementById("localVideo");
     v.style.webkitFilter="contrast(" + val + ")";
   };
   ```

   Here, `localVideo` is the ID property of the HTML video tag for the local video playback.

3. Navigate your web browser to the web page. You will first see an unprocessed video from the web camera. The following screenshot depicts such a situation:

4. On the left-hand side of the page, you can see a control described as **Contrast**. Try to move it to the right or left—you will see that the video has more and less contrast respectively. In the following screenshot I moved the control to the right and increased the contrast:

If you move it too much to the left, you will see just a light-gray box. If you move the control to the right, you will make the image almost black (depends on the amount of light at your place).

How it works...

When you move the control, the JavaScript function from the second step is called. This function applies the `contrast` filter with the control's value to the video object—using its style HTML property. From now on, the web browser will show this video applying the filter on the fly.

Working with saturation

In this recipe, we will cover the process of controlling the saturation of a video being captured from the web camera using WebRTC. Saturation is rarely used as a control available to users. Although for some kinds of applications it might be very useful.

How to do it...

Perform the following steps:

1. Add a control element to you application's main web page—using this object we will change the saturation's level:

   ```
   Saturation
   <input type="range"
   oninput="changeSaturation(this.valueAsNumber);"
   value="0" step="0.1" min="0" max="10">
   ```

2. Add an appropriate JavaScript function:

   ```
   function changeSaturation(val) {
     var v = document.getElementById("localVideo");
     v.style.webkitFilter="saturate(" + val + ")";
   };
   ```

   Here `localVideo` is the ID property of the HTML video tag for the local video playback.

3. Navigate your web browser to the web page. You will first see an unprocessed video from the web camera with normal saturation. The following screenshot depicts such a situation:

4. On the left-hand side of the page, you can see a control described as **Saturation**. Try to move it to the extreme left—you will see that the video became black and white. By smoothly moving the control to the right, you will add saturation, and the video will look more normal. In the following screenshot, I moved the control too much to the right, making the video too saturated:

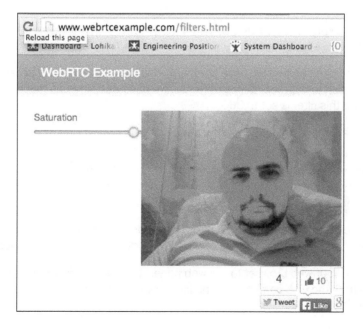

Moving the control to the right bound will make the video oversaturated, and it will be barely possible for us to see what's happening in the scene.

How it works...

When you move the control, the JavaScript function from the second step is called. This function applies the `saturate` filter using the control's value as the filter's parameter. The function uses the object's style HTML property. From now on, the web browser will show the video applying the chosen filter to it on the fly.

Working with hue

In this recipe, we will learn how to control the video's hue. Usually, you will not use this filter in your applications, although, sometimes it might be helpful; for example when you're using some kind of specific video equipment that might need this way of processing video.

How to do it...

Follow the given steps:

1. Add a control element to the application's main web page—using this object we will change the video's hue:

    ```
    Hue
    <input type="range"
    oninput="changeHue(this.valueAsNumber);"
    value="0" step="20" min="0" max="360">
    ```

    Here, you can see that we have set the max value as 360—this is because the hue's value is tied to degrees. In this universe, we have 360 degrees, so the maximum value for this filter is set to 360.

2. Add an appropriate JavaScript function:

    ```
    function changeHue(val) {
      var v = document.getElementById("localVideo");
      v.style.webkitFilter="hue-rotate(" + val + "deg)";
    };
    ```

    We have also added the `deg` postfix to the filter's value—it means degree. Here, `localVideo` is the ID property of the HTML video tag for the local video playback.

3. Navigate your web browser to the web page. You will first see an unprocessed video from the web camera, with no filter applied. The following screenshot depicts this stage:

4. On the left-hand side of the page, you can see a control described as **Hue**. Try to move it to the left and right—you will see that the video's colors change. This is because by moving the control, you change the image's hue. In the following screenshot I moved the control to the right, making the person's face dark pink, and the yellow-blue flag became white-green:

You probably will use this filter rarely. It can be useful if in case for some reason you have a broken video (from your web camera or from the peer) with abnormal hues. Otherwise, it can be applied just for fun.

How it works...

When you move the control, the JavaScript function from the second step is called. This function applies the `hue-rotate` filter using the control's value as the filter's degree. The function uses the object's style HTML property. From now on, the web browser will show the video applying the chosen filter to it on the fly.

Using the sepia filter

This recipe covers the usage of the sepia filter to process a video captured from a remote peer or local web camera using WebRTC. This is a popular filter often used as a special effect for making video applications more friendly and warm.

How to do it...

The following steps will show you how to use the sepia filter:

1. Add a control element to the main web page of the application you're developing—using this object we will control the value of the applied **Sepia** filter:

   ```
   Sepia
   <input type="range"
   oninput="changeSepia(this.valueAsNumber);"
   value="0" step="0.1" min="0" max="1">
   ```

2. Add an appropriate JavaScript function:

   ```
   function changeSepia(val) {
     var v = document.getElementById("localVideo");
     v.style.webkitFilter="sepia(" + val + ")";
   };
   ```

   Here, `localVideo` is the ID property of the HTML video tag for the local video playback.

3. Navigate your web browser to the web page. You will first see a raw video in the web camera, with no filter applied. In the following screenshot, you can see an image without any applied filter:

4. On the left-hand side of the page, you can see a control described as **Sepia**. Try to move it to the left and right—you will see that the video's colors change. The leftmost position makes the image look normal (no filter is applied). The rightmost position applies the filter to the most available value. In the following screenshot, I moved the control to the rightmost end and made the video look as if it was taken from an old movie:

How it works...

When you move the control, the JavaScript function from the second step is called. This function applies the `sepia` filter to the video image. The function uses the object's style HTML property. From now on, the web browser will show the video applying the chosen filter to it on the fly.

Using the opacity filter

In this recipe, we will cover how to use the opacity filter. You will probably rarely use it, but it can be used for implementing interesting features, such as picture in picture.

How to do it...

Follow these steps:

1. Add a control element to the main web page of your application—using this object we will control the video's opacity:

   ```
   Opacity
   <input type="range"
   oninput="changeOpacity(this.valueAsNumber);"
   value="1" step="0.1" min="0" max="1">
   ```

2. Add an appropriate JavaScript function:

   ```
   function changeOpacity(val) {
     var v = document.getElementById("localVideo");
     v.style.webkitFilter="opacity(" + val + ")";
   };
   ```

   Here `localVideo` is the ID property of the HTML video tag for the local video playback.

3. Navigate your web browser to the web page. You will first see an unprocessed video from the web camera, with no filter applied. The following screenshot depicts this stage:

4. On the left-hand side of the page, you can see a control described as **Opacity**. Try to move it to the left and right—you will see that the video becomes less and more transparent, respectively. The top-right position is the normal state, and the top-left position is the transparent state. In the following screenshot I moved the control a little to the left, and you can see that the person in the image is barely visible because of the image's transparency:

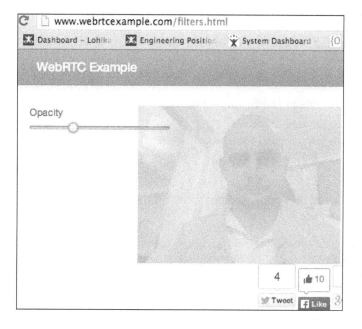

This filter can also be useful when you overlap several videos. Another utility of this filter is that, in case you're developing a multiuser conference, by using this filter and changing a users' video transparency, you can mark the participants as *currently speaking* or *on hold* accordingly.

How it works...

When you move the control, the JavaScript function from the second step is called. This function applies the `opacity` filter to the video using the control's value. The function uses the object's style HTML property. From now on, the web browser will show the video applying the chosen filter to it on the fly.

Inverting colors

This recipe covers the process of using a pretty simple filter: **inversion of colors**. It will hardly be useful for you in most normal cases, but it might be helpful if for some reason your peer sends you a broken video with inverted colors, or you get one from your web camera. Some cameras might work that way due to hardware incompatibility or due to the incorrect installation of software drivers.

How to do it...

Perform the following steps:

1. Add a control button to your application's main web page:

    ```
    Inversion
    <input type="range"
    oninput="invertColors(this.valueAsNumber);"
    value="0" step="0.1" min="0" max="1">
    ```

2. Add an appropriate JavaScript function:

    ```
    function invertColors(val) {
      var v = document.getElementById("localVideo");
      v.style.webkitFilter="invert(" + val + ")";
    };
    ```

    Here `localVideo` is the ID property of the HTML video tag for the local video playback.

3. Navigate your browser to the web page. You will first see an unprocessed video from the web camera. The following screenshot depicts this stage:

4. On the top left of the web page you can see the **Inversion** control. Try to move it to the left and right—you will see that the video image's colors change as and when you move the control to the left and right. In the following screenshot I moved the control almost to the rightmost position, and the image transformed to color negative of the original image:

How it works...

When you click on the **Inversion** button, the JavaScript function from the second step is called. This function applies the `invert` filter with the appropriate value to the video object—using its style HTML property. From now on, the web browser will show this video applying the filter online.

Implementing the blur effect

This recipe dives into the implementation of the blur effect. If you have worked on graphic editing computer software, then you are likely familiar with this effect.

How to do it...

The following steps will help you understand how to implement the blur effect:

1. Add a control element to the index web page of your application—using this object we will control the blur effect:

```
Blur
<input type="range"
oninput="doBlur(this.valueAsNumber);"
value="0" step="1" min="0" max="15">
```

2. Add an appropriate JavaScript function:

```
function doBlur(val) {
  var v = document.getElementById("localVideo");
  v.style.webkitFilter="blur(" + val + "px)";
};
```

We have added a `px` postfix for the filter's value—this is because of the blur's intensity that is setting in pixels. Here, `localVideo` is the ID property of the HTML video tag for the local video playback.

3. Navigate your web browser to the web page. You will first see the raw, unprocessed video from the web camera, with no filter applied. The following screenshot depicts this stage:

4. In the preceding screenshot, on the left-hand side of the page, you can see a **Blur** control. By moving this control to the left and right, you can set the intensity of the blurriness in an image. The leftmost position means that there is no blur and you should see a normal image. In the following screenshot I moved the control a little to the right from the middle, and you can see that the image became very blurry—it is barely possible to recognize the person in the video:

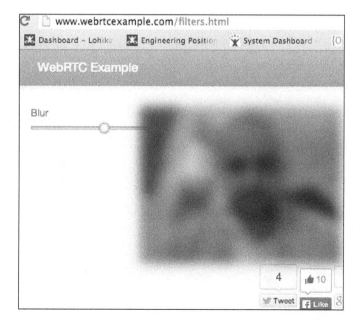

This filter can be used for indicating that you have muted someone on the videoconference, or for indicating that the conference has not started.

How it works...

When you move the control, the JavaScript function from the second step is called. This function applies the `blur` filter to the video using the control's value. The function uses the object's style HTML property. From now on, the web browser will show the video applying the chosen filter on the fly.

Implementing the dropped shadow effect

In this recipe, we will cover the process of implementing the dropped shadow effect. This filter can be used for decoration purposes. Although it utilizes CPU resources very actively, don't put it on every page.

How to do it...

Follow these steps:

1. Add a control element to the appropriate web page of the application—using this object we will control intensity of the effect:

```
Shadow
<input type="range"
oninput="doShadow(this.valueAsNumber);"
value="0" step="5" min="0" max="50">
```

2. Add the `onLoad` handler to HTML's `body` tag of the web page. By using this method, we will initialize the dropped shadow effect.

```
<body onload="doShadow(0);">
```

3. Add an appropriate JavaScript function:

```
function doShadow(val) {
  var v = document.getElementById("localVideo");
  v.style.webkitFilter="drop-shadow(" + val + "px " + val +
  "px 10px green)";
};
```

We have added the `px` postfix for the filter's value—this is because of the effect's intensity is setting in pixels. Also, you can see that we have set the shadow's width to 10 pixels, and we want the shadow to be green.

Here, `localVideo` is the ID property of the HTML video tag for the local video playback.

4. Navigate your web browser to the web page. You should see a normal image from your web camera, but there should be a green shadow around the image. The following screenshot depicts this stage:

5. On the left-hand side of the page you can see the **Shadow** control. By moving this control to the left and right, you can control the shadow's position. The leftmost position of the control is the initial position of the shadow—just around the image with a width of 10 pixels, as we have set it. In the following screenshot I moved the control a little to the right from the middle, and you can see that the green shadow has also moved to the bottom and right:

This filter can be used for additional UI decoration while developing WebRTC applications. You can easily control the shadow's size, position, and color.

How it works...

When you move the control, the JavaScript function from the second step is called. This function applies the `drop-shadow` filter to the video using the control's value. The function uses the object's style HTML property. From now on, the web browser will show the video applying the chosen filter on the fly.

Combining filters

All the filters described in this chapter can be combined and work together. In this recipe, we will cover this topic using a simple practical example—combining two filters: brightness and contrast.

How to do it...

Follow the given steps:

1. Add two control objects to the page for each of the filters we plan to use:

```
Brightness<br>
<input type="range" oninput="doFilter('brightness',
this.valueAsNumber);" value="0" step="0.1" min="0"
max="10">
<br>
Contrast<br>
<input type="range" oninput="doFilter('contrast',
this.valueAsNumber);" value="0" step="0.1" min="0"
max="10">
<br>
```

2. Add a global variable where we will store the values for each filter:

```
var filters = {};
```

3. Add an appropriate JavaScript function that will be called when value of the controls (introduced in the step 1) is changing:

```
function doFilter(filtername, val) {
  filters[filtername] = val;
  var v = document.getElementById("localVideo");
  var f = "";
  for (var fname in filters) {
    f = " " + fname + "(" + filters[fname] + ")" + f; }
  v.style.webkitFilter=f;
};
```

4. Navigate your web browser to the web page. You should see the usual image from your web camera and two controls: for brightness and contrast. By moving these controls, you can change the value of the image's contrast and brightness. Changing the value of one filter doesn't reset the value of another. In the following screenshot, you can see such a web page with the described feature:

How it works...

In the `doFilter` function, we get the name of a certain filter as the `filtername` parameter. Certain filter names we get there from the appropriate filter's control (refer to the first step). As the second parameter of the function, we also get certain filter values we have to use applying the filter.

After getting the filter name and filter value in the function, we will store these parameters (refer to the second step) in the `filters` array variable (we will use it as an associative array). Then we will go through all the array keys and values (filter names and their values) and will construct the string `f`, combining necessary filter names and its values. We will delimit filters by the space symbol.

After that, we will get the `f` string as something like the following:

```
brightness(3) contrast(5)
```

We will use the constructed string to change the style of the appropriate video tag. As a result, we will apply two filters in parallel.

You can combine as many filters as you like, but you should know that some of them could be resource hungry. If your application sets too many filters at one time, it might cause issues (the web browser might stack, for example).

Custom video processing

Until now, we considered standard filters only. In this recipe, we will cover the basic case of custom video processing. Using that approach, you can implement your own filters and processing algorithms.

How to do it...

As an example, we will implement the pixelization effect.

1. Put a canvas object somewhere on the application's web page. This canvas will be used for getting frames from the video. The visibility option is set to `hidden`—we don't want to show this canvas to the user, we will use it for our internal, technical purposes only.

   ```
   <canvas id="canva" width="384px" height="288px"
   style="visibility:hidden;"></canvas>
   ```

2. Put another canvas object on the web page. This canvas will be used to show the result of the video processing:

   ```
   <canvas id="fcanva" width="384px" height="288px"></canvas>
   ```

3. Add a button, which will enable the processing:

   ```
   <button onclick="pixelize(10)">Pixelize</button><br>
   ```

4. Implement the `pixelize` function. This function actually performs all the video processing:

   ```
   var pixelsize = 10;
   var w = 384;
   var h = 288;

   function pixelize(pixelsize) {
       cnv.drawImage(lv, 0, 0, w, h);
       for(var x = 1; x < w; x += pixelsize)
       {
           for(var y = 1; y < h; y += pixelsize)
           {
               var pxl = cnv.getImageData(x, y, 1, 1);
               fcnv.fillStyle =
               "rgb("+pxl.data[0]+","+pxl.data[1]+",
               "+pxl.data[2]+")";
   ```

```
            fcnv.fillRect(x, y, x + pixelsize - 1, y +
            pixelsize - 1);
        }
    }

    setTimeout(function () {
        pixelize(pixelsize);
    }, 0);
}
```

5. In the following screenshot, you can see how the filter works. On the left-hand side, the original video is shown, on the right-hand side, you can see the same video after applying the custom filter:

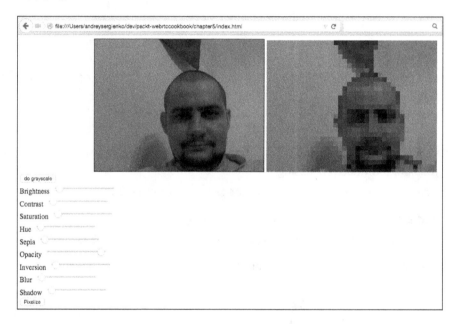

How it works...

We used two canvases: one (which is hidden) was used to copy frames from the video stream and get pixels; the second canvas was used to show processed video frames.

When the `pixelize` function is called the first time, it completes processing of the first video frame and then sets up a timer to be called the next time. Thus, the browser calls this function again and again. With every call, it gets a new video frame, processes it, and gets displayed using the second canvas object.

That way, you can implement any video frame processing algorithm and use it as your custom video filter.

6
Native Applications

In this chapter, we will cover the following topics:

- ▶ Building a customized WebRTC demo for iOS
- ▶ Compiling and running an original demo for iOS
- ▶ Compiling and running a demo for Android
- ▶ Building an OpenWebRTC library

Introduction

This chapter is fully dedicated to using WebRTC technology while developing native applications for mobile platforms. Here, the term **native application** refers to the kind of software that is being developed using native tools and SDK of a certain mobile platform.

First of all, you will learn how to get and compile WebRTC libraries that can be used for developing native applications. There is no separate code for every certain platform. Basically, the code base is the same for all available mobile platforms.

In other recipes, we will build and run WebRTC demo applications for Android and iOS, to demonstrate the use of WebRTC on mobile devices.

The *Building a customized WebRTC demo for iOS* recipe covers customized demo applications. The problem is that the WebRTC code base is under active development, and original example applications might not demonstrate all available features of the technology. For example, the original iOS example didn't support video calls for a long time and supported audio calls only. Nevertheless, it is possible to build a native iOS application that supports WebRTC video calls, and the custom demo application demonstrates that.

Software development for mobile platforms is a very specific field. It is barely possible to cover development of an application in just one chapter. So I assume that you have enough experience of developing software for certain mobile platforms, because this is something that is out of this book's topic. Here, we will only cover WebRTC specific details and skip the rest.

The flow of building a native application using WebRTC might seem tricky and non-trivial. The following diagram represents the general case with the basic steps of the flow:

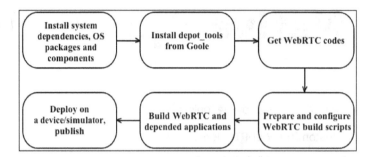

In this chapter, we will cover this flow with all its steps. We will also learn how we can make this process easier and simpler.

Building a customized WebRTC demo for iOS

In this recipe, we will download a simple, prepared WebRTC native demo application for iOS, compile it, and run it on a real device. This application can be used for video conference calls via Google's demo website, `https://apprtc.webrtc.org`.

This demo software is customized, meaning that WebRTC libraries are precompiled and should be just linked during compilation of the demo application. It also contains some changes compared to the original demo from Google.

Getting ready

The demo application is supposed to run on a device, not in a simulator. So you should be prepared with a physical Apple device (iPhone, iPad) to work on this recipe.

You should be registered on the iOS Developer Program by Apple to be able to install the application on your device. If you're not participating in this program, it is worth considering joining. For details, please refer to the program's official web page at `http://developer.apple.com`.

In my case, I used the following tools:

- iPhone 5s with iOS 8.0.2.
- A notebook with Windows 7 installed as the second device to build the WebRTC communication channel.
- In the notebook, I used a Chrome browser to run a WebRTC application.
- Xcode 6 to compile the iOS demo. For Xcode, you also need to have an OS X machine that runs.

How to do it...

Perform the following steps to build a customized WebRTC demo:

1. Create a new project directory and go to it as follows:

   ```
   mkdir ~/dev
   cd ~/dev
   ```

2. Get the source code using the following command:

   ```
   git clone https://github.com/fycth/webrtc-ios
   ```

3. Open the demo project in Xcode: `~/dev/webrtc-ios/ios-example/AppRTCDemo.xcodeproj`.

4. Choose the build target using the Xcode menu by navigating to **Product | Destination | iPhone**.

5. Build the demo application by navigating to **Product | Build**.

6. Connect your iPhone to the machine and run the demo by navigating to **Product | Run**.

After the last command is executed, the demo application will be installed on the device and will start automatically; it can take a couple of seconds, so don't rush to run the application manually.

In the following screenshot, you can see an icon of the installed **AppRTCDemo** application:

After the application starts, you will see a short message and a prompt to enter a room number. Navigate your browser on another machine to `http://apprtc.webrtc.org`; you will see an image from your camera. Copy the room number from the URL string and enter it in the demo application. The following screenshot represents this stage:

After you enter the code and click on the **Apply** button, the application will try to connect to the virtual room. It can take a couple of seconds (even up to one minute in my case), so be patient.

When the connection is established, you should see the image from the iPhone in the web browser, and vice versa. The following screenshot depicts a screenshot from my iPhone after I established a WebRTC connection with a notebook:

In the screenshot, you can see a man with an iPhone, from which this screenshot was taken. The video on the iPhone is translated from the notebook's camera. And the following screenshot represents what was visible on the notebook's display:

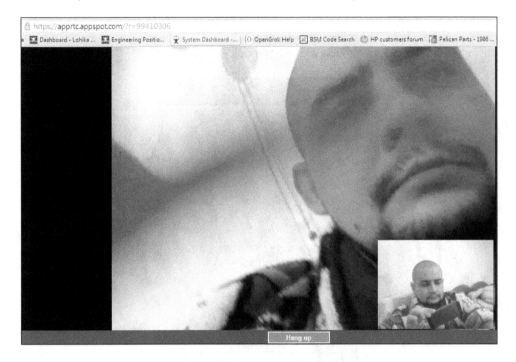

Here, in the small image box you can see the video taken from the notebook's web camera. In the big image, you can see the video translated from the iPhone.

There's more...

For this recipe I forked the code from another project on GitHub. To learn more, refer to `https://github.com/gandg/webrtc-ios`.

I introduced some changes in the forked project, fixing some minor issues. You can fork any of these projects and take its code as the base of your own project.

You can also check this project at `https://github.com/pristineio/webrtc-build-scripts`. It is a set of scripts developed specially to facilitate the compilation of WebRTC libraries' code for iOS. If you develop WebRTC-based software for Apple mobile OS, this tool might be very useful for you.

Building a demo project for a iOS simulator

This demo project uses precompiled WebRTC libraries that are built to use on physical devices. You should rebuild these libraries in case you want to run the application under an iOS simulator.

1. Download and install Google Developer Tools:

   ```
   mkdir ~/dev

   cd ~/dev

   git clone https://chromium.googlesource.com/chromium/tools/depot_
   tools.git

   export PATH=`pwd`/depot_tools:"$PATH"
   ```

2. Configure the developer tools:

   ```
   gclient config http://webrtc.googlecode.com/svn/trunk
   ```

3. Inform the tools that we want to build libraries for iOS:

   ```
   echo "target_os = ['ios']" >> .gclient
   ```

4. Download the WebRTC source code. It can take a couple of minutes; it will download several gigabytes of code.

   ```
   gclient sync
   ```

5. Configure the build tool as follows:

   ```
   export GYP_DEFINES="build_with_libjingle=1 build_with_chromium=0
   libjingle_objc=1"

   export GYP_GENERATORS="ninja"

   export GYP_DEFINES="$GYP_DEFINES OS=ios target_arch=ia32"

   export GYP_GENERATOR_FLAGS="$GYP_GENERATOR_FLAGS output_dir=out_
   sim"

   export GYP_CROSSCOMPILE=1

   gclient runhooks
   ```

6. Build the libraries as shown in the following command lines:

   ```
   cd ~/dev/trunk

   ninja -C out_sim/Debug iossim AppRTCDemo
   ```

   The building process can take some time. After that you will find compiled WebRTC libraries by navigating to ~/dev/trunk/out-sim/Debug/.

7. Now you should copy these libraries into the project's ios-example/libs folder, and then you will be able to build the project for iOS simulator.

See also

▶ Another recipe, *Building an OpenWebRTC library,* also might be useful for you in the scope of developing WebRTC native applications for iOS

▶ Refer to the *Compiling and running an original demo for iOS* recipe for details on how to work with the original demo from Google

Compiling and running an original demo for iOS

This recipe covers how to build an original Google WebRTC native demo application for iOS. The original demo from Google doesn't have any Xcode project files using which you could open the IDE and do the job with comfort. Unfortunately, you would have to use a set of console tools and scripts to compile this application.

Getting ready

In this recipe, we will cover the process of building an application for both an iOS simulator and for a physical device. So you should have a Mac OS X machine to run the demo in a simulator, and you should have an Apple gadget if you would like to run it on a physical device.

You should also be registered on the Apple iOS Developer Program to be able to install your application on your device. If you're not participating in this program, it is worth considering joining. For details, refer to the program's official web page at `http://developer.apple.com`.

In my case, I used a MacBook Pro with Mac OS X 10.9.5 installed on it.

How to do it...

First of all, we need to download and build the WebRTC source code. The demo application is a part of this code, so we will build it with the rest by performing the following steps:

1. Download and install Google Developer Tools:

   ```
   mkdir -p ~/dev && cd ~/dev
   git clone https://chromium.googlesource.com/chromium/tools/depot_
   tools.git
   export PATH=`pwd`/depot_tools:"$PATH"
   ```

2. Configure the developer tools:

   ```
   gclient config http://webrtc.googlecode.com/svn/trunk
   ```

3. Inform the tools that we want to build libraries for iOS:

```
echo "target_os = ['ios','mac']" >> .gclient
```

4. Download the WebRTC source code. It can take a couple of minutes; it will download several gigabytes of code:

```
gclient sync
```

Building a demo project for an iOS device

The following steps should be taken if you're building a demo to run on a physical Apple device. If you want to run the demo on an iOS simulator, skip this section and continue to the next one:

1. Configure the build tool as follows:

```
export GYP_DEFINES="build_with_libjingle=1 build_with_chromium=0
libjingle_objc=1"

export GYP_GENERATORS="ninja"

export GYP_DEFINES="$GYP_DEFINES OS=ios target_arch=armv7"

export GYP_GENERATOR_FLAGS="$GYP_GENERATOR_FLAGS output_dir=out_
ios"

export GYP_CROSSCOMPILE=1
```

2. Prepare the build scripts:

```
gclient runhooks
```

3. Build the demo application:

```
cd ~/dev/trunk

ninja -C out_ios/Debug-iphoneos AppRTCDemo
```

Building a demo project for an iOS simulator

This section describes the steps that should be taken if you want to compile the application for an iOS simulator. If you want to run the application on a physical device, find the relevant steps provided in the previous section:

1. Configure the build tool as follows:

```
export GYP_DEFINES="build_with_libjingle=1 build_with_chromium=0
libjingle_objc=1"

export GYP_GENERATORS="ninja"

export GYP_DEFINES="$GYP_DEFINES OS=ios target_arch=ia32"

export GYP_GENERATOR_FLAGS="$GYP_GENERATOR_FLAGS output_dir=out_
sim"

export GYP_CROSSCOMPILE=1
```

2. Prepare build scripts:

```
gclient runhooks
```

3. Build a demo application:

```
cd ~/dev/trunk
ninja -C out_sim/Debug iossim AppRTCDemo
```

4. Start the application in an iOS simulator:

```
~/dev/trunk/out_sim/Debug/AppRTCDemo.app
```

There's more...

The original code from Google doesn't have any IDE project files so you have to deal with console scripts through all the development process. This can be easier if you use some third-party tools that simplify the building process. Such kinds of tools can be found at `http://tech.pristine.io/build-ios-apprtc/`.

See also

▶ It is also worth taking a look at the *Building a customized WebRTC demo for iOS* recipe. In this recipe we cover the process of using a ready-to-use Xcode simple project with precompiled WebRTC binaries.

Compiling and running a demo for Android

Here, you will learn how to build a native demo WebRTC application for Android. Unfortunately, the supplied demo application from Google doesn't contain any IDE-specific project files, so you will have to deal with console scripts and commands during all the building process.

Getting ready

We will need to check whether we have all the necessary libraries and packages installed on the work machine. For this recipe, I used a Linux box—Ubuntu 14.04.1 x64. So all the commands that might be specific for OS will be relevant to Ubuntu. Nevertheless, using Linux is not mandatory and you can take Windows or Mac OS X.

 If you're using Linux, it should be 64-bit based. Otherwise, you most likely won't be able to compile Android code.

Preparing the system

First of all, you need to install the necessary system packages:

```
sudo apt-get install git git-svn subversion g++ pkg-config gtk+-2.0
libnss3-dev libudev-dev ant gcc-multilib lib32z1 lib32stdc++6
```

Installing Oracle JDK

By default, Ubuntu is supplied with OpenJDK, but it is highly recommended that you install an Oracle JDK. Otherwise, you can face issues while building WebRTC applications for Android. One another thing that you should keep in mind is that you should probably use Oracle JDK version 1.6—other versions (in particular, 1.7 and 1.8) might not be compatible with the WebRTC code base. This will probably be fixed in the future, but in my case, only Oracle JDK 1.6 was able to build the demo successfully.

1. Download the Oracle JDK from its home page at `http://www.oracle.com/technetwork/java/javase/downloads/index.html`.

   In case there is no download link on such an old JDK, you can try another URL: `http://www.oracle.com/technetwork/java/javasebusiness/downloads/java-archive-downloads-javase6-419409.html`.

    Oracle will probably ask you to sign in or register first. You will be able to download anything from their archive.

2. Install the downloaded JDK:

   ```
   sudo mkdir -p /usr/lib/jvm
   cd /usr/lib/jvm && sudo /bin/sh ~/jdk-6u45-linux-x64.bin
   --noregister
   ```

   Here, I assume that you downloaded the JDK package into the home directory.

3. Register the JDK in the system:

   ```
   sudo update-alternatives --install /usr/bin/javac javac /usr/lib/jvm/jdk1.6.0_45/bin/javac 50000
   sudo update-alternatives --install /usr/bin/java java /usr/lib/jvm/jdk1.6.0_45/bin/java 50000
   sudo update-alternatives --config javac
   sudo update-alternatives --config java
   cd /usr/lib
   sudo ln -s /usr/lib/jvm/jdk1.6.0_45 java-6-sun
   export JAVA_HOME=/usr/lib/jvm/jdk1.6.0_45/
   ```

4. Test the Java version:

```
java -version
```

You should see something like Java HotSpot on the screen—it means that the correct JVM is installed.

Getting the WebRTC source code

Perform the following steps to get the WebRTC source code:

1. Download and prepare Google Developer Tools:

```
mkdir -p ~/dev && cd ~/dev
git clone https://chromium.googlesource.com/chromium/tools/depot_
tools.git
export PATH=`pwd`/depot_tools:"$PATH"
```

2. Download the WebRTC source code:

```
gclient config http://webrtc.googlecode.com/svn/trunk
echo "target_os = ['android', 'unix']" >> .gclient
gclient sync
```

The last command can take a couple of minutes (actually, it depends on your Internet connection speed), as you will be downloading several gigabytes of source code.

Installing Android Developer Tools

To develop Android applications, you should have **Android Developer Tools** (**ADT**) installed. This SDK contains Android-specific libraries and tools that are necessary to build and develop native software for Android. Perform the following steps to install ADT:

1. Download ADT from its home page `http://developer.android.com/sdk/index.html#download`.

2. Unpack ADT to a folder:

```
cd ~/dev
unzip ~/adt-bundle-linux-x86_64-20140702.zip
```

3. Set up the `ANDROID_HOME` environment variable:

```
export ANDROID_HOME=`pwd`/adt-bundle-linux-x86_64-20140702/sdk
```

How to do it...

After you've prepared the environment and installed the necessary system components and packages, you can continue to build the demo application:

1. Prepare Android-specific build dependencies:

    ```
    cd ~/dev/trunk
    source ./build/android/envsetup.sh
    ```

2. Configure the build scripts:

    ```
    export GYP_DEFINES="$GYP_DEFINES build_with_libjingle=1 build_
    with_chromium=0 libjingle_java=1 OS=android"
    gclient runhooks
    ```

3. Build the WebRTC code with the demo application:

    ```
    ninja -C out/Debug -j 5 AppRTCDemo
    ```

After the last command, you can find the compiled Android packet with the demo application at `~/dev/trunk/out/Debug/AppRTCDemo-debug.apk`.

Running on the Android simulator

Follow these steps to run an application on the Android simulator:

1. Run Android SDK manager and install the necessary Android components:

    ```
    $ANDROID_HOME/tools/android sdk
    ```

    Choose at least Android 4.x—lower versions don't have WebRTC support. In the following screenshot, I've chosen Android SDK 4.4 and 4.2:

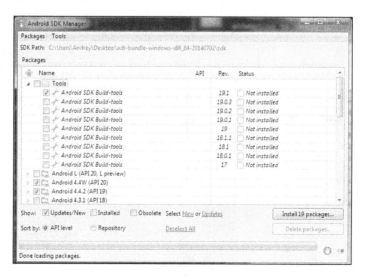

2. Create an Android virtual device:

```
cd $ANDROID_HOME/tools
```

```
./android avd &
```

The last command executes the Android SDK tool to create and maintain virtual devices. Create a new virtual device using this tool. You can see an example in the following screenshot:

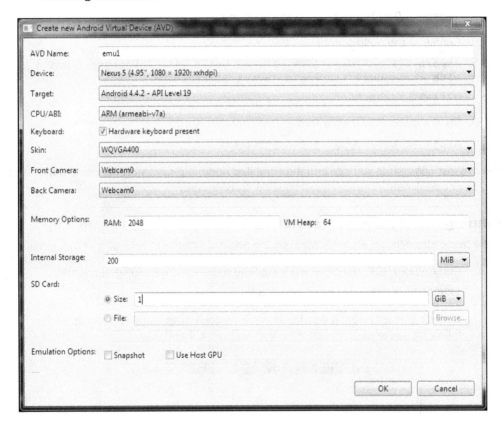

3. Start the emulator using just the created virtual device:

```
./emulator -avd emu1 &
```

This can take a couple of seconds (or even minutes), after that you should see a typical Android device home screen, like in the following screenshot:

4. Check whether the virtual device is simulated and running:

    ```
    cd $ANDROID_HOME/platform-tools
    ./adb devices
    ```

    You should see something like the following:

    ```
    List of devices attached
    emulator-5554    device
    ```

    This means that your just created virtual device is OK and running; so we can use it to test our demo application.

5. Install the demo application on the virtual device:

    ```
    ./adb install ~/dev/trunk/out/Debug/AppRTCDemo-debug.apk
    ```

    You should see something like the following:

    ```
    636 KB/s (2507985 bytes in 3.848s)
    pkg: /data/local/tmp/AppRTCDemo-debug.apk
    Success
    ```

    This means that the application is transferred to the virtual device and is ready to be started.

6. Switch to the simulator window; you should see the demo application's icon. Execute it like it is a real Android device. In the following screenshot, you can see the installed demo application AppRTC:

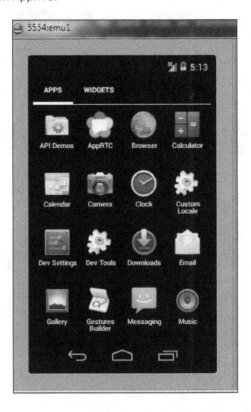

 While trying to launch the application, you might see an error message with a Java runtime exception referring to `GLSurfaceView`. In this case, you probably need to switch to the **Use Host GPU** option while creating the virtual device with **Android Virtual Device** (**AVD**) tool.

Fixing a bug with GLSurfaceView

Sometimes if you're using an Android simulator with a virtual device on the ARM architecture, you can be faced with an issue when the application says **No config chosen**, throws an exception, and exits.

This is a known defect in the Android WebRTC code and its status can be tracked at `https://code.google.com/p/android/issues/detail?id=43209`.

The following steps can help you fix this bug in the original demo application:

1. Go to the `~/dev/trunk/talk/examples/android/src/org/appspot/apprtc` folder and edit the `AppRTCDemoActivity.java` file. Look for the following line of code:

   ```
   vsv = new AppRTCGLView(this, displaySize);
   ```

2. Right after this line, add the following line of code:

   ```
   vsv.setEGLConfigChooser(8,8,8,8,16,16);
   ```

   You will need to recompile the application:

   cd ~/dev/trunk

   ninja -C out/Debug AppRTCDemo

3. Now you can deploy your application and the issue will not appear anymore.

Running on a physical Android device

For deploying applications on an Android device, you don't need to have any developer certificates (like in the case of iOS devices). So if you have an Android physical device, it probably would be easier to debug and run the demo application on the device rather than on the simulator.

1. Connect the Android device to the machine using a USB cable.
2. On the Android device, switch the USB debug mode on.
3. Check whether your machine sees your device:

   cd $ANDROID_HOME/platform-tools

   ./adb devices

   If device is connected and the machine sees it, you should see the device's name in the result print of the preceding command:

   List of devices attached

   Q04721C35410   device

4. Deploy the application onto the device:

   cd $ANDROID_HOME/platform-tools

   ./adb -d install ~/dev/trunk/out/Debug/AppRTCDemo-debug.apk

   You will get the following output:

   3016 KB/s (2508031 bytes in 0.812s)

   pkg: /data/local/tmp/AppRTCDemo-debug.apk

   Success

After that you should see the **AppRTC** demo application's icon on the device.

After you have started the application, you should see a prompt to enter a room number. At this stage, go to `http://apprtc.webrtc.org` in your web browser on another machine; you will see an image from your camera. Copy the room number from the URL string and enter it in the demo application on the Android device. Your Android device and another machine will try to establish a peer-to-peer connection, and might take some time. In the following screenshot, you can see the image on the desktop after the connection with Android smartphone has been established:

Here, the big image represents what is translated from the frontal camera of the Android smartphone; the small image depicts the image from the notebook's web camera. So both the devices have established direct connection and translate audio and video to each other.

The following screenshot represents what was seen on the Android device:

There's more...

The original demo doesn't contain any ready-to-use IDE project files; so you have to deal with console commands and scripts during all the development process. You can make your life a bit easier if you use some third-party tools that simplify the building process. Such tools can be found at `http://tech.pristine.io/build-android-apprtc`.

See also

▶ If you consider developing WebRTC applications for iOS, the *Building a customized WebRTC demo for iOS* recipe might also be useful for you

Building an OpenWebRTC library

At the beginning of 2014, Ericsson presented its own open source implementation of WebRTC stack—OpenWebRTC. Ericsson states that this product supports iOS, Android, Windows, Linux, and Mac OS X platforms from the box. In this recipe, we will build this new WebRTC stack. This implementation came out just a couple of days ago and there isn't a ready-to-use example supplied with it, so we will build just the library.

Getting ready

At this time, OpenWebRTC build scripts support Linux and Mac OS X platforms only, and there is no ready solution to build OpenWebRTC under Windows. So you need Linux or OS X installed to work on this recipe.

In my case, I used a Mac Book Pro with Mac OS X 10.9 installed.

How to do it...

Perform the following steps to build OpenWebRTC:

1. Get the source codes:

   ```
   mkdir ~/dev && cd ~/dev
   git clone git@github.com:EricssonResearch/openwebrtc.git
   --recursive
   cd openwebrtc
   ```

2. Configure the environment (this step will take some time). If you're working under Linux, put `linux` instead of `osx` in the command:

   ```
   cd scripts/bootstrap
   ./bootstrap.sh -r osx
   cd -
   ```

3. Build the dependencies. You can also use `linux` and `android` words if you're building for the appropriate platforms. Note that you need Android NDK installed and configured to build dependencies for this platform:

   ```
   cd scripts/dependencies
   ./build-all.sh -r osx ios
   ./deploy_deps.sh
   cd -
   ```

4. Build OpenWebRTC using the following command:

   ```
   ./build.sh -r osx ios
   ```

After all these commands are executed, you will have OpenWebRTC libraries built and ready to use. To further learn this library, it might be worth taking a look at Bowser—an open sourced web browser completely built on the OpenWebRTC stack.

There's more...

This new library is under active development and even its documentation actively changes. So, for more details, please refer to the home page of the project at `http://www.openwebrtc.io`.

Also take a look at Bowser—an open source WebRTC-oriented web browser from Ericsson. This browser can run under both Android and iOS. Its home page is at `http://www.openwebrtc.io/bowser/`.

7
Third-party Libraries

In this chapter, we will cover the following topics:

- ▶ Building a video conference using SimpleWebRTC
- ▶ Creating an application using RTCMultiConnection
- ▶ Developing a simple WebRTC chat using PeerJS
- ▶ Making a simple video chat with rtc.io
- ▶ Using OpenTok to create a WebRTC application
- ▶ Creating a multiuser conference using WebRTCO

Introduction

When a new technology or an instrument appears on the market, it might not be reasonable to create your own framework or a library by utilizing this new tool to develop a product. Sometimes it is worth looking around and using a **Software Development Kit** (**SDK**) or a ready-to-use framework that implements all the technology's necessary features.

WebRTC is a very young technology that is under active development. We don't have a completed standard yet, only a draft. There are many third-party frameworks and libraries available that utilize WebRTC features and provide a nice API for a developer. To use such tools, it is not necessary to get deep into WebRTC and standards, but you can concentrate just on your product.

 Most of the frameworks provide you with a complete set of tools. Therefore, you might need to use Google's adapter.js in addition to keep compatibility between multiple web browsers or their (browsers') versions.

Usually, such an SDK can make a developer's life easier—they often provide additional services such as signaling and STUN/TURN servers. When using a good third-party framework, you often don't need to take care of the server infrastructure and installation and maintenance of your own signaling server; you can work only on the client code—the rest will be served by the chosen solution.

In this chapter, we will consider a few such tools. You will find recipes that utilize a tool's API to implement basic examples of WebRTC applications. All examples are based on the official tool's documentation and demo applications from their home pages.

 WebRTC stack is developed with great attention to security, and the web browser might not even run the application in case it is accessed from the local system. So while testing the provided examples, place them on a web server. As an alternative, you can use cloud services such as Dropbox for accessing the application over public folder—in this case, you should change all HTTP links in the application to HTTPS.

Building a video conference using SimpleWebRTC

SimpleWebRTC is a very easy-to-use framework written in JavaScript. Using this product, you can start your first video conference in just one minute. In this recipe, we will cover the process of creating of a basic WebRTC application using the SimpleWebRTC software.

Getting ready

In this recipe, we will create a simple HTML page by utilizing a SimpleWebRTC framework. So, you will need a text editor and a WebRTC compliant web browser. If you're using Firefox, the demo might be executed from the local filesystem; if you're using Chrome, you should use a web server—otherwise, the browser will prohibit the running of the application.

How to do it...

To build a basic videoconference using this tool, you need to create just one HTML web page. You don't even need to register an account in the vendor's system.

1. Create an empty HTML file and add the following code:

```
<!DOCTYPE html>
<html>
<head lang="en">
<meta charset="UTF-8">
```

2. Include a SimpleWebRTC JavaScript framework:

```
<script src="http://simplewebrtc.com/latest.js"></script>
</head>
<body>
```

3. Create a video object for a local video:

```
<video height="300" id="localVideo"></video>
```

4. Create a video object for the video translated from a remote peer:

```
<div id="remotesVideos"></div>
<br>
```

5. Create a button and tie a handler function to it. When you click the button, videoconference will be created:

```
<button id="btn1" onclick="startconf()">Start
conference</button>
<script language="JavaScript">
```

6. Set up a variable to handle the `SimpleWebRTC` object:

```
var webrtc = null;
```

The following function is called when a customer clicks the button:

```
function startconf() {
```

7. Create a `SimpleWebRTC` object with initial parameters. We will send the IDs of both the video objects (for local and remote video); also, we will ask the framework to get media access immediately:

```
webrtc = new SimpleWebRTC({
    localVideoEl: 'localVideo',
    remoteVideosEl: 'remotesVideos',
    autoRequestMedia: true
});
```

The following code actually starts the videoconference. Here, we will also set up a virtual room name, `Room86#`—you are free to use any name you would like to use:

```
webrtc.on('readyToCall', function () {
    webrtc.joinRoom('Room86#');
});
};
</script>
</body>
</html>
```

8. Now, save this file in a folder and open it in your web browser (in my case, I've used Firefox for Mac OS X).

How it works...

When you open the HTML file in your web browser, you will see a blank page with a button. Click on the **Start conference** button—the web browser will capture a video from your web camera and show it on the page (it may ask you for access permission).

In the following screenshot, you can see this stage:

Now, it is time to connect another peer. Open the same HTML file in another browser. You can even copy it to another machine and open it there. Then click on the **Start conference** button—after a couple of seconds, the peer connection should be established and you should see both the local and remote images on every browser window, as shown in the following screenshot:

Note you don't need to install a signaling server—SimpleWebRTC takes care of it. When you call SimpleWebRTC's JavaScript API methods, it communicates to the signaler server installed on the SimpleWebRTC's servers.

There's more...

Although we considered a very simple example of using a SimpleWebRTC framework, this tool can be used to build more complex applications. For more details, please refer to the official documentation for the framework at `http://simplewebrtc.com`.

Creating an application using RTCMultiConnection

This recipe covers the process of creating a simple WebRTC application using an open source RTCMultiConnection framework. This is a JavaScript-based framework that allows you to build applications and services using many WebRTC features, including experimental features.

Getting ready

To work with this framework, we will build a basic WebRTC service that supports private virtual rooms for videoconferencing. You will need to write some HTML and JavaScript code, which does not need to develop any server-side parts. So, having just a text editor and a WebRTC compliant web browser should be enough to work on this recipe.

How to do it...

The RTCMultiConnection tool takes all of the work regarding the signaling on its own. Thus, you can concentrate on the client side and UI.

1. Create an empty HTML file and add the following code inside it:

```
<!DOCTYPE html>
<html>
<head lang="en">
    <meta charset="UTF-8">
```

2. Include the HTML style supplied with the tool. This is not necessary, and you can use your own CSS:

```
<link rel="stylesheet"
href="http://cdn.webrtc-experiment.com/style.css">
```

3. Include the framework's JavaScript libraries:

```
<script src="http://cdn.webrtc-
experiment.com/firebase.js"> </script>
<script src="http://cdn.webrtc-
experiment.com/RTCMultiConnection.js">
</script></head>
```

```
<body>
<section>
<span>
```

4. The following anchor is used for creating virtual rooms:

```
<a href="" target="_blank" title=""><code>
<strong id="unique-token"></strong></code></a></span>
```

5. Add an input object to handle the virtual room name:

```
<input type="text" id="conference-name">
```

6. Create a new button element on the page. When it is clicked, a new conference will start:

```
<button id="setup-new-conference" class="setup">Setup
New Conference</button>
</section>
<table style="width: 100%;" id="rooms-list"></table>
```

7. Create a separate `div` layer for the video objects:

```
<div id="videos-container"></div>
</section>
<script>
```

8. Create a new connection object. Using this object, we can control the connection itself:

```
var connection = new RTCMultiConnection();
connection.session = {
    audio: true,
    video: true
};
```

9. Declare a callback handler that will be called when a new media stream is ready. This handler will create a new video object for every media stream and place it in the video container layer:

```
connection.onstream = function(e) {
    e.mediaElement.width = 300;
    videosContainer.insertBefore(e.mediaElement,
    videosContainer.firstChild);
};
```

10. Create a handler for the stream ended event. It will be called when a stream is stopped (peer connection is interrupted, for example). This function will remove the irrelevant video object:

```
connection.onstreamended = function(e) {
    e.mediaElement.style.opacity = 0;
    setTimeout(function() {
```

```
            if (e.mediaElement.parentNode) {
                e.mediaElement.parentNode.
                removeChild(e.mediaElement);
            }
        }, 1000);
    };
    var sessions = { };
```

11. Make a function that will be called when a new virtual room is created, and someone is waiting for the remote peer to join:

```
connection.onNewSession = function(session) {
    if (sessions[session.sessionid]) return;
    sessions[session.sessionid] = session;
    var tr = document.createElement('tr');
```

12. We need to notify the customer when the virtual room is created. The following code shows such a notification and creates a **Join** button:

```
tr.innerHTML = '<td><strong>' +
session.extra['session-name'] + '</strong> is
running a conference!</td>' + '<td><button
class="join">Join</button></td>';
roomsList.insertBefore(tr, roomsList.firstChild);
var joinRoomButton = tr.querySelector('.join');
joinRoomButton.setAttribute('data-sessionid',
session.sessionid);
```

13. Create an appropriate code for the **Join** button:

```
joinRoomButton.onclick = function() {
    this.disabled = true;
    var sessionid = this.getAttribute('data-
    sessionid');
    session = sessions[sessionid];
    if (!session) throw 'No such session exists.';
    connection.join(session);
};
};
var videosContainer = document.getElementById('videos-
container') || document.body;
    var roomsList = document.getElementById('rooms-list');
    document.getElementById('setup-new-conference').onclick
    = function() {
        this.disabled = true;
        connection.extra = {
        'session-name': document.getElementById
        ('conference-name').value || 'Anonymous'
```

```
    };
    connection.open();
  };
connection.connect();
```

14. The unique URL to share the virtual room with others is created on the client side as well. The following code represents how this task is solved in the example:

```
(function() {
    var uniqueToken = document.getElementById('unique-
    token');
    if (uniqueToken)
        if (location.hash.length > 2)
          uniqueToken.parentNode.parentNode.parentNode.
          innerHTML = '<h2 style="text-align:center;">
          <a href="' + location.href + '"
          target="_blank">Share this link</a></h2>';
        else uniqueToken.innerHTML =
          uniqueToken.parentNode.parentNode.href = '#'
          + (Math.random() * new
          Date().getTime()).toString(36).toUpperCase().
          replace( /\./g , '-');
})();
</script>
</body>
</html>
```

That is all. Save this file on a disk, and navigate your web browser to it.

How it works...

When you open the HTML file, you will see a web page similar to the following:

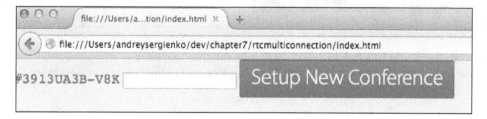

Now, create a new private virtual room by clicking on the URL to the left (it will open a new tab in the browser as shown in the following screenshot).

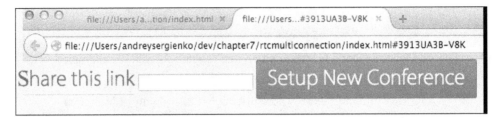

In this page, you should enter your name or a room's name in the input textbox, and then click on the **Setup New Conference** button. After that, you should see the image from your web camera:

Now, copy the **Share this link** URL and open it on another machine, or you can open it in another browser's tab, like I did. You will see a big **Join** button like the one shown in the following screenshot:

So, to connect to the conference, just click on the **Join** button. Right after that, the conference will try to establish peer-to-peer connection. If everything goes well, every peer should see both local and remote images.

 In my case, I used the same machine (just separate browser windows), so the images are identical.

This library uses Firebase (`https://www.firebase.com`) for signaling, so you don't need to install and maintain your own signaling server—RTCMultiConnection will take care of that.

There's more...

RTCMultiConnection allows you to create more complex applications, and utilize advanced WebRTC features. Here, we touched just the basic concepts.

For details on how to use this framework, refer to its official home page `https://www.webrtc-experiment.com/RTCMultiConnection/`.

Developing a simple WebRTC chat using PeerJS

In this recipe, we will use the PeerJS WebRTC framework to create a simple web chat concept by utilizing data channels.

Getting ready

PeerJS requires developers to register before they can use its API. During the registration process (it is free), a developer gets a unique ID that can be used to work with the API. If you would like to use this framework and don't mind registering, then visit its home page at `http://peerjs.com`.

How to do it...

Using PeerJS is really simple, and a basic example can be performed using just one HTML file. In the following steps, you will find such an index file with comments in all the important places:

1. Place the standard HTML headers:

   ```
   <!DOCTYPE html>
   <html>
   <head lang="en">
       <meta charset="UTF-8">
   ```

2. Include the PeerJS library:

   ```
   <script
   src="http://cdn.peerjs.com/0.3/peer.js"></script>
   </head>
   <body>
   ```

3. Add an input textbox. Here, a customer can enter his/her name while connecting to the system. For simplicity, the same box will be used to enter further chat messages:

```
<br><input type="text" id="inputbox"/>
```

4. Create three buttons to connect to the system, to call the remote peer, and to send messages to the remote peer:

```
<button id="btn_connect"
onclick="Connect()">Connect!</button>
<button id="btn_call" onclick="CallTo()"
disabled="true">Call To</button>
<button id="btn_send" onclick="SendMessage()">Send
message</button>
<script language="JavaScript">
```

5. In the following variable, you should add your developer API ID you got from the PeerJS system during the registration process:

```
var MY_API_ID = YOUR_API_ID;
    var peer = null;
    var conn = null;
```

6. The following function takes the customer's name and registers it in the PeerJS system. After that, another peer can connect to this customer using its name for connection:

```
function Connect() {
    var myname =
    document.getElementById("inputbox").value;
    peer = new Peer(myname, {key: MY_API_ID});
```

7. Set up a callback function on the connection event. This function will be called when a remote peer establishes a connection with us. Here, we will also set a helper function that will print the received messages from the remote peer to the browser's console:

```
peer.on('connection', function(connection) {
    connection.on('data', function(data){
        console.log("Remote peer said: " + data);
    });
    conn = connection;
});
document.getElementById("btn_connect").
setAttribute("disabled", "true");
document.getElementById("btn_call").
removeAttribute("disabled");
};
```

8. We also need a function that will call the remote peer. The following code represents such a function. It takes a remote peer's name from the input textbox and calls PeerJS to establish the connection:

```
function CallTo() {
    var remotename =
    document.getElementById("inputbox").value;
    conn = peer.connect(remotename);
    document.getElementById("btn_call").
    setAttribute("disabled","true");
};
```

9. To send messages, we need an appropriate function responsible for that. Such a function, you can find in the following:

```
function SendMessage() {
    var msg =
    document.getElementById("inputbox").value;
    conn.send(msg);
};
</script>
</body>
</html>
```

That is all. Save this file on a disk, and navigate your web browser to the demo.

How it works...

Open a prepared HTML file, in that, you will see an input box and three buttons. Enter a peer's name in the textbox and click on the **Connect!** button. It will connect to a PeerJS system. Now, open the file in another browser (we can also open the file on another machine). Enter another peer's name, and click on **Connect!**. In the following screenshot, I used `peer1` and `peer2` as names for the peers:

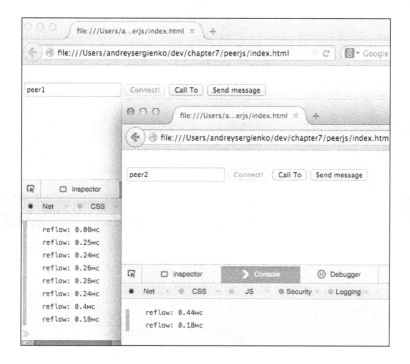

Now, for the second peer, enter the first peer's name (peer1 in my case) in the textbox, and click on the **Call To** button. This will start to establish the peer connection—peer2 will try to make a call to peer1.

After the connection is established, we can test message exchanging. For peer2, enter any input in the textbox and click on **Send message**. The entered text will be sent to peer1, and will be printed in its browser console. In the following screenshot, I have sent a Hello, peer1!!! message:

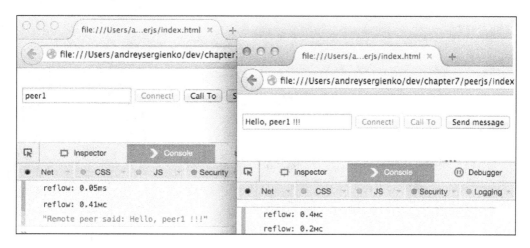

PeerJS uses its own infrastructure such as signaling mechanisms. Thus, if you use PeerJS, you don't need to be worried of developing signaling protocols, and you can concentrate on developing your application.

There's more...

You can find out more about PeerJS from its home page `http://peerjs.com`.

This is a free and open source tool, so it can be used as an SDK or can be taken as a code base for developing another WebRTC framework for custom application.

Making a simple video chat with rtc.io

rtc.io is a free and open source project for developing WebRTC applications. It provides simple and clean APIs. In this recipe, we will use rtc.io to create a basic video chat service.

Getting ready

Like most of the other considered frameworks, rtc.io serves its own signaling server, so you can create a basic application using just a few lines of JavaScript code and HTML. For this recipe, you will need a text editor and web browser.

How to do it...

Create an empty file in the text editor and add the following code. This is a plain HTML with a JavaScript section. Relevant places are commented inline.

1. First of all, let's add the standard HTML heads and bit of styles:

```
<!DOCTYPE html>
<html>
<head lang="en">
    <meta charset="UTF-8">
    <style>
        #messages {
            border: 1px solid black;
            min-height: 20px;
        }
    </style>
```

2. Include the rtc.io framework in this project:

```
<script src="https://rawgit.com/rtc-
io/rtc/master/dist/rtc.js">
</script></head>
<body>
```

3. Create separate `div` layers for chat messages and both local and remote video:

```
<div id="messages" contenteditable></div>
<div id="l-video"></div>
<div id="r-video"></div>
<script language="JavaScript">
```

4. Set the framework's options—for a basic case, we just need a room's name and signaler server URL. Here, we used a native signaler sever hosted on the rtc.io infrastructure. It's an open source code, so you can download and install it on your own server:

```
var rtcOpts = {
    room: 'my-cool-test-room',
    signaller: '//switchboard.rtc.io'
};
```

5. Initialize the framework and create an RTC object:

```
var rtc = RTC(rtcOpts);
var localVideo = document.getElementById('l-video');
var remoteVideo = document.getElementById('r-video');
var messageWindow =
document.getElementById('messages');
```

6. Bind handler functions to appropriate events that might be generated on the data channel:

```
function bindDataChannelEvents(id, channel, attributes,
connection) {
    channel.onmessage = function (evt) {
        messageWindow.innerHTML = evt.data;
    };
    messageWindow.onkeyup = function () {
        channel.send(this.innerHTML);
    };
}
```

7. Initialize the session:

```
function init(session) {
    session.createDataChannel('chat');
    session.on('channel:opened:chat',
    bindDataChannelEvents);
}
```

8. Display the local and remote video:

```
localVideo.appendChild(rtc.local);
remoteVideo.appendChild(rtc.remote);
```

9. Handle the session establishing event:

```
        rtc.on('ready', init);
    </script>
</body>
</html>
```

The example can be saved on the disk and uploaded to the web server.

How it works...

We created a new RTC object using the framework's API. Additionally, we set a couple of functions to handle events. Then, we initialized the framework by calling the appropriate API method. After all this, it will handle signaling and peer connections.

There's more...

For additional details and advanced examples of how to use this framework, refer to its homepage at `http://rtc.io`.

Using OpenTok to create a WebRTC application

OpenTok is a proprietary framework that allows you to build WebRTC-based applications using the provided SDK. In this recipe, we will build a simple demo application by utilizing the basic features of the tool.

Getting ready

To use this framework, you should register with the OpenTok system, and get a unique developer API ID. To use this system, you should have three keys: the API key, session ID, and token. The following instructions cover the process of creating these keys:

1. Navigate to `https://tokbox.com/opentok/` and click on **Sign Up**.

2. Fill the form and click on the **Sign Up** button:

3. Check for an e-mail from OpenTok (TokBox), they will send a confirmation e-mail with the API key. Write down the API key—this is the first key. Confirm your registration with their system by clicking on the appropriate link in the e-mail:

4. Navigate to `https://dashboard.tokbox.com`—find the **Projects** section and click on the **View Details** button:

5. In the next page, you will see **Project Tools**, where you can create a new session. Do it by using the **Create** button:

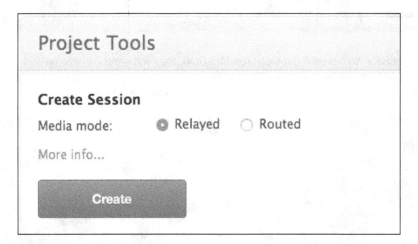

6. Right after that, you will see the generated session ID below the button. Write down this value—this is the second key.

7. After you've created the new session, you should create a new token based on this session. At the **Generate Token** section, click on the **Generate** button:

8. After you've clicked on the **Generate** button, you will see a generated token below the button, as shown in the following screenshot:

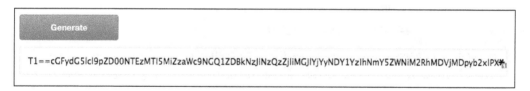

9. Write down this value (generated token)—*this is the third key*.

Now, you have all the three keys to work with the OpenTok system.

 This framework requires you to use a web server, so for this recipe, you should have a web server installed and configured.

How to do it...

Now, when you have your API ID, a session ID, and two tokens, you can continue with the process of building an application using OpenTok:

1. Create an empty HTML file (let's name it `index.html`) and add the following code:

```
<!DOCTYPE html>
<html>
<head lang="en">
    <meta charset="UTF-8">
    <title></title>
```

2. Include the OpenTok code in your project:

```
<script src =
'http://static.opentok.com/webrtc/v2.2/js/opentok.min.js'>
</script>
    <script type="text/javascript">
```

3. In the following lines, you should insert the actual API ID (API key) and session ID, which you have generated while preparing for this recipe:

```
var apiKey = <YOUR_API_ADI>;
var sessionId = <GENERATED_SESSION_ID>;
```

4. As you remember, we generated two tokens—one per client. Add the first token in the following variable:

```
var token = <TOKEN_1>;
```

5. Initialize the session by calling the OpenTok method:

```
var session = OT.initSession(apiKey, sessionId);
```

6. Subscribe to events:

```
session.on("streamCreated", function(event) {
    session.subscribe(event.stream);
});
```

7. Open a new connection:

```
session.connect(token, function(error) {
    var publisher = OT.initPublisher();
    session.publish(publisher);
});
    </script>
</head>
<body>
```

8. We also need an HTML object to publish a video there:

```
<h1>Awesome video feed!</h1>
<div id="myPublisherDiv"></div>
</body>
</html>
```

9. Now, save the file and create another one (let's name it `index2.html`). Make the second file identical to the first one. Then, edit the second file and change the token value in the following line:

```
var token = <TOKEN_2>;
```

10. In the second file, you should add the second token, which you have generated while preparing for this recipe. Save the second file.

Now, we have two files: `index.html` and `index2.html`. They are both identical, except for their token value—every file contains its own token ID. Put both the files in the web server.

How it works...

It's time to test what we've developed.

Open a web browser and navigate to the place where the first file (`index.html`) is located. Note that you should not use a filesystem, and both the files should be accessible on the web server. After the page is opened, you will see an image from the web camera.

Now, on another machine, open a web browser and navigate to the second file (index2. html). You will see the similar picture. In a couple of seconds, the connection will be established and you will see local and remote images on both the machines. The following screenshot represents this case:

In my case, I have used the same machine, but opened the files in two different web browsers.

OpenTok takes care of signaling and other technical processes. As you can see, the application is very compact, and the code is very short and clean. You don't need to spend time on installation and maintenance of server components, they are provided and transparently served by the framework.

There's more...

We considered just a simple example of using OpenTok, but this tool allows you to create more complex applications with advanced features. For details, refer to OpenTok's home page at https://tokbox.com/opentok/.

Creating a multiuser conference using WebRTCO

In this recipe, we will create a simple application that supports a multiuser videoconference. We will do it using WebRTCO—an open source JavaScript framework for developing WebRTC applications.

Getting ready

For this recipe, you should have a web server installed and configured. The application we will create can work while running on the local filesystem, but it is more convenient to use it via the web server.

To create the application, we will use the signaling server located on the framework's homepage. The framework is open source, so you can download the signaling server from GitHub and install it locally on your machine. GitHub's page for the project can be found at `https://github.com/Oslikas/WebRTCO`.

How to do it...

The following recipe is built on the framework's infrastructure. We will use the framework's signaling server. What we need to do is include the framework's code and do some initialization procedure:

1. Create an HTML file and add common HTML heads:

    ```
    <!DOCTYPE html>
    <html lang="en">
    <head>
        <meta charset="utf-8">
    ```

2. Add some style definitions to make the web page looking nicer:

    ```
    <style type="text/css">
        video {
            width: 384px;
            height: 288px;
            border: 1px solid black;
            text-align: center;
        }
        .container {
            width: 780px;
            margin: 0 auto;
        }
    </style>
    ```

3. Include the framework in your project:

```
<script type="text/javascript" src =
"https://cdn.oslikas.com/js/WebRTCO-1.0.0-beta-min.js"
charset="utf-8"></script>
</head>
```

4. Define the `onLoad` function—it will be called after the web page is loaded. In this function, we will make some preliminary initializing work:

```
<body onload="onLoad();">
```

5. Define HTML containers where the local video will be placed:

```
<div class="container">
    <video id="localVideo"></video>
</div>
```

6. Define a place where the remote video will be added. Note that we don't create HTML video objects, and we just define a separate div. Further, video objects will be created and added to the page by the framework automatically:

```
<div class="container" id="remoteVideos"></div>
<div class="container">
```

7. Create the controls for the chat area:

```
<div id="chat_area" style="width:100%; height:250px;
overflow: auto; margin:0 auto 0 auto; border:1px solid
rgb(200,200,200); background: rgb(250,250,250);"></div>
</div>
<div class="container" id="div_chat_input">
    <input type="text" class="search-query"
    placeholder="chat here" name="msgline" id="chat_input">
    <input type="submit" class="btn" id="chat_submit_btn"
    onclick="sendChatTxt();"/>
</div>
```

8. Initialize a few variables:

```
<script type="text/javascript">
    var videoCount = 0;
    var webrtco = null;
    var parent = document.getElementById('remoteVideos');
    var chatArea = document.getElementById("chat_area");
    var chatColorLocal = "#468847";
    var chatColorRemote = "#3a87ad";
```

9. Define a function that will be called by the framework when a new remote peer is connected. This function creates a new video object and puts it on the page:

```
function getRemoteVideo(remPid) {
    var video = document.createElement('video');
    var id = 'remoteVideo_' + remPid;
    video.setAttribute('id',id);
    parent.appendChild(video);
    return video;
}
```

10. Create the `onLoad` function. It initializes some variables and resizes the controls on the web page. Note that this is not mandatory, and we do it just to make the demo page look nicer:

```
function onLoad() {
    var divChatInput =
    document.getElementById("div_chat_input");
    var divChatInputWidth = divChatInput.offsetWidth;
    var chatSubmitButton =
    document.getElementById("chat_submit_btn");
    var chatSubmitButtonWidth =
    chatSubmitButton.offsetWidth;
    var chatInput =
    document.getElementById("chat_input");
    var chatInputWidth = divChatInputWidth -
    chatSubmitButtonWidth - 40;
    chatInput.setAttribute("style","width:" +
    chatInputWidth + "px");
    chatInput.style.width = chatInputWidth + 'px';
    var lv = document.getElementById("localVideo");
```

11. Create a new WebRTCO object and start the application. After this point, the framework will start signaling connection, get access to the user's media, and will be ready for income connections from remote peers:

```
webrtco = new WebRTCO('wss://www.webrtcexample.com/signalling',
lv, OnRoomReceived, onChatMsgReceived, getRemoteVideo, OnBye);
};
```

Here, the first parameter of the function is the URL of the signaling server. In this example, we used the signaling server provided by the framework. However, you can install your own signaling server and use an appropriate URL. The second parameter is the local video object ID. Then, we will supply functions to process messages of received room, received message, and received remote video stream. The last parameter is the function that will be called when some of the remote peers have been disconnected.

12. The following function will be called when the remote peer has closed the connection. It will remove video objects that became outdated:

```
function OnBye(pid) {
    var video = document.getElementById("remoteVideo_"
    + pid);
    if (null !== video) video.remove();
};
```

13. We also need a function that will create a URL to share with other peers in order to make them able to connect to the virtual room. The following piece of code represents such a function:

```
function OnRoomReceived(room) {
    addChatTxt("Now, if somebody wants to join you,
    should use this link: <a
    href=\""+window.location.href+"?
    room="+room+"\">"+window.location.href+"?
    room="+room+"</a>",chatColorRemote);
};
```

14. The following function prints some text in the chat area. We will also use it to print the URL to share with remote peers:

```
function addChatTxt(msg, msgColor) {
    var txt = "<font color=" + msgColor + ">" +
    getTime() + msg + "</font><br/>";
    chatArea.innerHTML = chatArea.innerHTML + txt;
    chatArea.scrollTop = chatArea.scrollHeight;
};
```

15. The next function is a callback that is called by the framework when a peer has sent us a message. This function will print the message in the chat area:

```
function onChatMsgReceived(msg) {
    addChatTxt(msg, chatColorRemote);
};
```

16. To send messages to remote peers, we will create another function, which is represented in the following code:

```
function sendChatTxt() {
    var msgline =
    document.getElementById("chat_input");
    var msg = msgline.value;
    addChatTxt(msg, chatColorLocal);
    msgline.value = '';
    webrtco.API_sendPutChatMsg(msg);
};
```

17. We also want to print the time while printing messages; so we have a special function that formats time data appropriately:

```
function getTime() {
    var d = new Date();
    var c_h = d.getHours();
    var c_m = d.getMinutes();
    var c_s = d.getSeconds();

    if (c_h < 10) { c_h = "0" + c_h; }
    if (c_m < 10) { c_m = "0" + c_m; }
    if (c_s < 10) { c_s = "0" + c_s; }
    return c_h + ":" + c_m + ":" + c_s + ": ";
};
```

18. We have some helper code to make our life easier. We will use it while removing obsolete video objects after remote peers are disconnected:

```
Element.prototype.remove = function() {
    this.parentElement.removeChild(this);
}
NodeList.prototype.remove =
HTMLCollection.prototype.remove = function() {
    for(var i = 0, len = this.length; i < len; i++) {
        if(this[i] && this[i].parentElement) {
            this[i].parentElement.removeChild(this[i]);
        }
    }
}
</script>
</body>
</html>
```

Now, save the file and put it on the web server, where it could be accessible from web browser.

How it works...

Open a web browser and navigate to the place where the file is located on the web server. You will see an image from the web camera and a chat area beneath it. At this stage, the application has created the WebRTCO object and initiated the signaling connection. If everything is good, you will see an URL in the chat area. Open this URL in a new browser window or on another machine—the framework will create a new video object for every new peer and will add it to the web page.

The number of peers is not limited by the application. In the following screenshot, I have used three peers: two web browser windows on the same machine and a notebook as the third peer:

Using this framework, you can attain your own signaling server or you can use the one that is provided by the tool.

There's more...

For now, the tool supports basic WebRTC features and it is in the beta stage. WebRTCO is under development and it might be improved in the future.

For details on this framework, refer to its home page at `http://www.oslikas.com/webrtco`.

Source codes and examples can be found on the GitHub page at `https://github.com/Oslikas/WebRTCO`.

More examples can be found on the demo page, `http://www.webrtcexample.com`.

8
Advanced Functions

In this chapter, we will cover the following topics:

- ▸ Visualizing a microphone's sound level
- ▸ Muting a microphone
- ▸ Pausing a video
- ▸ Taking a screenshot
- ▸ Streaming media

Introduction

This chapter covers advanced examples of using WebRTC features. The following recipes allow you to improve your application's usability and make it friendlier by adding advanced features and functionality.

All the recipes in this chapter are oriented on the client side and implemented in JavaScript. Some of them appear to be pretty simple and others might be more complex, but the main purpose of these recipes is to make the application more adaptable to real life and friendly for customers.

Visualizing a microphone's sound level

If your application works with audio and video (for example, you're developing a video conferencing service), it would be probably a good idea to add a live indication of the microphone sound level. Using this feature, peers can estimate and control their microphone's audio levels. So, in this recipe, we're implementing microphone activity indication.

Getting ready

This recipe is simple, and you will just need a text editor to create and edit HTML. To test this recipe, you should have a web server installed and configured—it is highly recommended to test the example via a web server rather than just on a local filesystem; otherwise, the web browser might block calls to the WebRTC API.

How to do it...

Perform the following steps:

1. Create an HTML file and insert the following codes. Note that the important places are commented inline:

```
<!DOCTYPE html>
<html>
<head lang="en">
    <meta charset="UTF-8">
```

2. Include the WebRTC adapter from Google. This file allows you to use universal function names in all supported web browsers:

```
<script src = "https://rawgit.com/GoogleChrome/webrtc/
master/samples/web/js/adapter.js"> </script>
</head>
<body>
```

3. Create a simple HTML element to display a local video from a web camera:

```
<div><video width="384" id="lVideo" muted="true"
autoplay="true"></video></div>
```

4. Create a canvas element—here, we will represent the microphone's sound level:

```
<canvas width="384" height="20" id="micecanvas"
style="background-color: white;"></canvas>
<script type="text/javascript">
```

The following function gets access to user media:

```
function init() {
    var constraints = {"audio": true, "video":
    {"mandatory": {}, "optional": []}};
    getUserMedia(constraints, onUserMediaSuccess,
    onUserMediaError);
}
```

5. We need to handle errors, so a simple error handler function can be found in the following code:

```
function onUserMediaError(error) {
    console.log("Error: " + error);
}
```

The following callback function will be called after the application has access to the user media:

```
function onUserMediaSuccess(stream) {
```

6. To attach a media stream to the video control, use the following code:

```
var localVideo =
document.getElementById("lVideo");
attachMediaStream(localVideo, stream);
```

7. Set up a function alias to make this work under different supported browsers:

```
window.AudioContext = window.AudioContext ||
window.webkitAudioContext ||
window.mozAudioContext;
```

8. Initialize the local variables and get access to the microphone:

```
var audioContext = new AudioContext();
var analyser = audioContext.createAnalyser();
var microphone =
audioContext.createMediaStreamSource(stream);
```

9. Assign a script processor to the audio context. By using the script processor, we will be able to process audio data and calculate microphone activity level:

```
var javascriptNode =
audioContext.createScriptProcessor(2048, 1, 1);
analyser.smoothingTimeConstant = 0.3;
analyser.fftSize = 1024;
microphone.connect(analyser);
analyser.connect(javascriptNode);
javascriptNode.connect
(audioContext.destination);
var canvasContext =
document.getElementById("micecanvas").
getContext("2d");
```

10. Set up an audio data processing function—here, we will do all the calculations:

```
javascriptNode.onaudioprocess = function() {
    var array =  new
    Uint8Array(analyser.frequencyBinCount);
    analyser.getByteFrequencyData(array);
```

```
var values = 0;
var length = array.length;
for (var i = 0; i < length; i++) {
    values += array[i];
}
```

11. Calculate the average sound level value and draw it on the canvas:

```
var average = values / length;
canvasContext.clearRect(0, 0, 384, 20);
canvasContext.fillStyle = 'red';
canvasContext.fillRect(0, 0, average, 20);
    }
}
```

12. The following function starts the application:

```
    init();
</script>
</body>
</html>
```

13. Now, save the file and put it on the web server, making it accessible through a certain URL.

14. Navigate to the URL. You will see the image from your web camera, and a short horizontal red bar beneath it. You will just see the local video because we haven't implemented an interconnection with remote peers.

15. Now, talk through the microphone and make some noise—the bar will respond to the sound by changing its length and trembling. This bar represents the microphone's sound activity level and you can estimate it visually.

How it works...

Using WebRTC API, we will create the audio context and audio analyzer objects. Then, we will get access to the microphone. We will also create `ScriptProcessor` with a buffer of 2048 bytes, and one input and one output channel. Using the `fftSize` attribute of the analyzer, we will set the size of the **Fast Fourier Transform** (**FFT**) buffer to 1024. We will connect the analyzer and the script processor, and then, we will set up the `onaudioprocess` handler function. Now, approximately every 0.3 seconds, we will get a signal from the browser to our handler function where we use received data to calculate the sound volume and to draw it on the bar.

See also

▶ Regarding detailed explanations of the possible usage of the audio API, you can refer to its official documentation at http://webaudio.github.io/web-audio-api/

Muting a microphone

Usually, voice calling software has a microphone muting feature. So, you can enable or disable your microphone during the call, deciding whether the remote peer should hear your voice or not. In this recipe, we will implement such a feature for a WebRTC application.

Getting ready

For this example, you don't need any preliminary specific steps. Use your development environment as you usually do.

How to do it...

Follow these steps:

1. For this feature, you need to add a button element to your HTML page. This button will enable or disable the microphone:

    ```
    <button id="mute_btn" onclick="muteBtnClick()">Mute
    Mic</button>
    ```

2. You also need to set up a handler for the `onclick` event of the element—it will do the actual work. The following code is an example of such a handler:

    ```
    function muteBtnClick() {
    ```

3. We will update our button with the microphone state, so we need to get the button ID:

    ```
        var btn = document.getElementById("mute_btn");
    ```

4. Before we can decide whether we want to mute or unmute the microphone, we should be able to know its actual state—for this purpose, we will use the `isMicMuted` function:

    ```
        if (isMicMuted()) {
    ```

5. Our microphone is muted, so we want to unmute it and update the button with the appropriate state:

    ```
            muteMic(false);
            btn.innerHTML = "Mute Mic";
        } else {
    ```

6. The microphone is unmuted, so we will mute it and update the button as well:

    ```
            muteMic(true);
            btn.innerHTML = "Unmute Mic";
        }
    }
    ```

7. In the handler, we will use the `isMicMuted` function to detect whether the microphone is muted. Let's implement this function as well:

```
function isMicMuted() {
        return !(localStream.getAudioTracks()[0].enabled); }
```

8. Note that the WebRTC API can let us know whether the audio track is enabled, but our function returns the microphone's muted value. So, we will invert the enabled value received from WebRTC stack.

9. Finally, we need to implement the actual mute/unmute function:

```
function muteMic (mute) {
        localStream.getAudioTracks()[0].enabled = !mute;
};
```

10. Here, `localStream` is a variable that contains a local stream object received after a successful call of the `getUserMedia` WebRTC API function.

 In this function, we will set up the `enabled` value, but the function gets the *should I mute the microphone* parameter. If this function gets true as an argument, it should set false to the `enabled` property of the audio track. This is why we will invert the value again, as we do it in the `isMicMuted` function.

How it works...

The main idea is to get an appropriate audio track of the local media stream, and to change its state to disabled or enabled. In the first case, the track will be muted and the remote peer will not hear your voice. Changing the state can be done in real time.

There's more...

If you have more than one audio device, the `getAudioTracks` function might return several audio tracks and it might be necessary to go over all of them:

```
var audiotracks = localStream.getAudioTracks();
for (var i = 0, l = audiotracks.length; i < l; i++)
{
        audiotracks[i].enabled = false;
}
```

See also

▸ Refer to the *Pausing a video* recipe to see a similar technique applied to video streams

Pausing a video

If you're participating in a video conference call, you might want to temporarily switch your video camera off and take a pause. During this time, your remote peer shouldn't see an image from your camera. In most videoconferencing software, you can enable or disable your camera during the call. In this recipe, we will implement this feature for a WebRTC application.

Getting ready

For this recipe, you don't need any specific preparations. Just create a basic conferencing WebRTC application.

How to do it...

Perform the following steps:

1. We need to add a **Pause Video** button to the application's web page:

   ```
   <button id="pause_video_btn"
   onclick="pauseVideoBtnClick()">Pause Video</button>
   ```

2. You also should set up a handler for the `onclick` event of the button:

   ```
   function pauseVideoBtnClick() {
   ```

3. We will update our button with the video stream state (whether it is paused or not), so we need to get the button ID:

   ```
       var btn = document.getElementById("pause_video_btn");
   ```

4. Before we decide whether we should pause the stream or start playing it back again, we should be able to know its current state—for this purpose, we will use the `isVideoPaused` function:

   ```
       if (isVideoPaused()) {
   ```

5. If the video stream is paused, we want to start playing it back and update the button with the new state, then use the following code:

   ```
           pauseVideo(false);
           btn.innerHTML = "Pause Video";
       } else {
   ```

6. In case the video is streaming, we will pause it and update the button as well:

```
pauseVideo(true);
btn.innerHTML = "Stream Video";
}
}
```

7. In the handler, we will use the `isVideoPaused` function to detect whether the video stream is paused. Let's implement this function as well:

```
function isVideoPaused() {
    return !(localStream.getVideoTracks()[0].enabled);
}
```

8. Note that the WebRTC API can let us know if a certain video track is enabled or not, but our function returns the *is the video paused* state. So, we will invert the enabled value received from the WebRTC stack.

9. Finally, we need to implement the function that actually puts the video on pause and vice versa:

```
function pauseVideo (pause) {
    localStream.getVideoTracks()[0].enabled = !pause;
};
```

10. Here, `localStream` is a variable that contains a local stream object received after a successful call of the `getUserMedia` WebRTC API function.

In this function, we will set up the `enabled` value, but the function gets the *should I put the video on pause* parameter. So, if it gets true as an argument, it should set the `enabled` property of the video track to false.

How it works...

The root idea of the described solution is to get an appropriate video track of the local media stream and to change its state to disabled or enabled. In the first case, the video track will be paused, streaming will be stopped, and the remote peer will not see you. Changing the state can be done in real time.

See also

▶ Refer to the *Muting a microphone* recipe for additional details regarding the usage of this solution to work with audio tracks

Taking a screenshot

Sometimes, it can be useful to be able to take screenshots from a video during videoconferencing. In this recipe, we will implement such a feature.

Getting ready

No specific preparation is necessary for this recipe. You can take any basic WebRTC videoconferencing application. We will add some code to the HTML and JavaScript parts of the application.

How to do it...

Follow these steps:

1. First of all, add image and canvas objects to the web page of the application. We will use these objects to take screenshots and display them on the page:

    ```
    <img id="localScreenshot" src="">
    <canvas style="display:none;" id="localCanvas"></canvas>
    ```

2. Next, you have to add a button to the web page. After clicking on this button, the appropriate function will be called to take the screenshot from the local stream video:

    ```
    <button onclick="btn_screenshot()" id="btn_screenshot">Make
    a screenshot</button>
    ```

3. Finally, we need to implement the screenshot taking function:

    ```
    function btn_screenshot() {
    var v = document.getElementById("localVideo");
    var s = document.getElementById("localScreenshot");
    var c = document.getElementById("localCanvas");
    var ctx = c.getContext("2d");
    ```

4. Draw an image on the canvas object—the image will be taken from the video object:

    ```
    ctx.drawImage(v,0,0);
    ```

5. Now, take reference of the canvas, convert it to the DataURL object, and insert the value into the src option of the image object. As a result, the image object will show us the taken screenshot:

    ```
    s.src = c.toDataURL('image/png');
    }
    ```

6. That is it. Save the file and open the application in a web browser. Now, when you click on the **Make a screenshot** button, you will see the screenshot in the appropriate image object on the web page. You can save the screenshot to the disk using right-click and the pop-up menu.

How it works...

We use the canvas object to take a frame of the video object. Then, we will convert the canvas' data to `DataURL` and assign this value to the `src` parameter of the image object. After that, an image object is referred to the video frame, which is stored in the canvas.

See also

▶ Refer to the *Visualizing a microphone's sound level* and *Muting a microphone* recipes for examples regarding how to work with audio data

Streaming media

This recipe covers another interesting feature that can be implemented using the WebRTC stack: streaming prerecorded media from one peer to another one.

Getting ready

We will stream a prerecorded WebM file, so you need to have one. You can download demo WebM files from the Internet. For example, from `http://www.webmfiles.org/demo-files/`.

In this recipe, we will create two files: an HTML page and a JavaScript library.

 This feature doesn't work on the local filesystem. To implement this feature, you need to have a web server where you can place all the application files, and where the application is accessible to the customer.

A signaling server is also necessary for this recipe. You can use the server from *Chapter 1, Peer Connections*.

How to do it...

Open your text editor, and let's create the HTML page by following the given steps:

1. Make a simple HTML header:

```
<!DOCTYPE html>
<html>
<head>
    <title>My WebRTC file media streaming demo</title>
```

2. Add some style for the video component:

```
<style type="text/css">
    video {
        width: 384px;
        height: 288px;
        border: 1px solid black;
        text-align: center;
    }
</style>
```

3. Include a JavaScript library that we will write at the next stage:

```
<script type="text/javascript"
src="myrtclib.js"></script>
```

4. Include Google's adapter to keep cross-browser compatibility:

```
<script
src="https://rawgit.com/GoogleChrome/webrtc/master
/samples/web/js/adapter.js"></script>
</head>
<body>
```

5. Create `div`, where a connection link will be published for peers:

```
<div id="status"></div><br>
```

6. Create a video element. This element will show the media streamed from the remote peer:

```
<div><video id="remotevideo" autoplay="true"
controls="true"></video></div>
```

7. Create a file choosing component and a button that will start the streaming process:

```
<div>
    file you want to stream <input type="file" id="files"
    name="files[]"/> then press <button
    onclick="onSendBtnClick()">Start streaming !</button>
</div>
<script>
    var filelist;
```

8. Check whether the web browser supports components and technologies that we use for this feature:

```
if (window.File && window.FileReader && window.FileList
&& window.Blob) {
    document.getElementById('files').
    addEventListener('change', handleFileSelect,
    false);
```

9. Connect to the signaling server and initialize our WebRTC library. Note that you should use an actual IP and port of the signaling server where it is running on your machine. By default, they are 127.0.0.1 and 30001, as implemented in the appropriate recipes of *Chapter 1*, *Peer Connections*, where we considered signaling servers:

```
myrtclibinit("ws://127.0.0.1:30001",
document.getElementById("remotevideo"));
} else {
```

10. Create an alert for instances when the web browser doesn't support necessary technologies:

```
alert('The File APIs are not fully supported in
this browser.');
}
```

11. Implement a function that handles the file choosing component:

```
function handleFileSelect(evt) {
    filelist = evt.target.files;
};
```

12. Implement a function that starts the streaming process. Note that the `doStreamMedia` function is implemented in the JavaScript library that will be considered in the next stage:

```
function onSendBtnClick() {
    doStreamMedia(filelist[0]);
};
```

13. Implement a callback function that constructs a connection link and publishes it on the web page:

```
function onRoomReceived(room) {
    var st = document.getElementById("status");
    st.innerHTML = "Now, if somebody wants to join you,
    should use this link: <a href=\""
    +window.location.href+"?room="+room+"\">"
    +window.location.href+"?room="+room+"</a>";
};
</script> </body> </html>
```

Next, you need to create a JavaScript library that is used in the HTML page we just created. Most of the code is simple and identical to the appropriate parts of the recipes from other chapters. Here, we will cover only specific moments that are important in the scope of the feature; known pieces of code will be skipped. Note that the full source code for this recipe is supplied along with this book.

This example actively uses WebRTC data channels, so you can refer to the *Implementing a chat using data channels* recipe from *Chapter 1, Peer Connections*, for more details on this topic. Perform the following steps for using data channels:

1. Declare a chunk size. While streaming the prerecorded media, the application reads the media file chunk by chunk and sends it to the remote peer. So, we have to declare the chunk size value—1024, in this particular case. You can play with other values and see how they affect the demo. Don't use too low or too high values:

```
var chunkSize = 1024;
```

2. Declare variables that will handle buffer and media source. The buffer is a structure that handles raw media data on the client side (where the media will be streamed). The media source represents a WebRTC object that will be tied with a video HTML object:

```
var receiverBuffer = null;
var recvMediaSource = null;
```

3. Declare a variable that will handle the HTML video object where the streamed media will be shown:

```
var remoteVideo = null;
```

4. Declare an array. This will be used as a cache to temporarily store the received chunks in case the remote peer sends them faster than we can draw them on the video:

```
var queue = [];
```

5. The following code is used for compatibility between Firefox and Chrome:

```
window.MediaSource = window.MediaSource ||
window.WebKitMediaSource;
```

6. Establish a new peer-to-peer data channel:

```
function createDataChannel(role) {
    try {
        data_channel =
        pc.createDataChannel("datachannel_"+room+role,
        null);
    } catch (e) {
        console.log('error creating data channel ' +
        e);
        return;
    }
    initDataChannel();
}
```

7. While setting a session description, remove bandwidth limitations. Some web browsers (for example, some versions of Chrome) limit bandwidth, so connection performance might degrade. To avoid that, we will call our custom `setBandwidth` function, which removes such limitations:

```
function setLocalAndSendMessage(sessionDescription) {
    sessionDescription.sdp =
    setBandwidth(sessionDescription.sdp);
    pc.setLocalDescription(sessionDescription,
    function() {}, failureCallback);
    sendMessage(sessionDescription);
};
```

8. Implement the `setBandwidth` function. It sets the bandwidth limit to a higher value instead of the default one, which might be set by the browser:

```
function setBandwidth(sdp) {
    sdp = sdp.replace( /a=mid:data\r\n/g ,
    'a=mid:data\r\nb=AS:1638400\r\n');
    return sdp;
}
```

9. Change the `onReceiveMessageCallback` function, adopting it for the new feature. You should be familiar with this function from *Chapter 1, Peer Connections*.

```
function onReceiveMessageCallback(event) {
    try {
        var msg = JSON.parse(event.data);
        if (msg.type === 'chunk') {
            onChunk(msg.data);
```

```
            }
        }
        catch (e) {}
    };
```

10. Declare the auxiliary variables for slicing the media file:

```
var streamBlob = null;
var streamIndex = 0;
var streamSize = 0;
```

11. Implement a function that is called from the HTML page. This function reads the media file, slices it into chunks, and sends them to the remote peer:

```
function doStreamMedia(fileName) {
    var fileReader = new window.FileReader();
    fileReader.onload = function (e) {
        streamBlob = new window.Blob([new
        window.Uint8Array(e.target.result)]);
        streamSize = streamBlob.size;
        streamIndex = 0;
        streamChunk();
    };
    fileReader.readAsArrayBuffer(fileName);
}
function streamChunk() {
    if (streamIndex >= streamSize)
    sendDataMessage({end: true});
    var fileReader = new window.FileReader();
    fileReader.onload = function (e) {
        var chunk = new
        window.Uint8Array(e.target.result);
        streamIndex += chunkSize;
        pushChunk(chunk);
        window.requestAnimationFrame(streamChunk);
    };
    fileReader.readAsArrayBuffer
    (streamBlob.slice(streamIndex,
    streamIndex + chunkSize));
}
```

12. Implement a function to receive media data. This function initializes the media source and buffer objects, and prepares to receive media chunks that are sent by the remote peer:

```
function doReceiveStreaming() {
    recvMediaSource = new MediaSource();
```

```
      remoteVideo.src =
      window.URL.createObjectURL(recvMediaSource);
      recvMediaSource.addEventListener('sourceopen',
      function (e) {
          remoteVideo.play();
```

13. We will use the WebM media file, so we should set an appropriate media type for the media buffer:

```
      receiverBuffer =
      recvMediaSource.addSourceBuffer
      ('video/webm; codecs="vorbis,vp8"');
      receiverBuffer.addEventListener('error',
      function(e) { console.log('error: ' +
      receiverBuffer.readyState); });
      receiverBuffer.addEventListener('abort',
      function(e) { console.log('abort: ' +
      receiverBuffer.readyState); });
      receiverBuffer.addEventListener('update',
      function(e) {
          if (queue.length > 0 &&
          !receiverBuffer.updating)
          doAppendStreamingData(queue.shift());
      });
      console.log('media source state: ',
      this.readyState);
      doAppendStreamingData(queue.shift());
  }, false);
      recvMediaSource.addEventListener('sourceended',
      function(e) { console.log('sourceended: ' +
      this.readyState); });
      recvMediaSource.addEventListener('sourceclose',
      function(e) { console.log('sourceclose: ' +
      this.readyState); });
      recvMediaSource.addEventListener('error',
      function(e) { console.log('error: ' +
      this.readyState); });
  };
```

14. The following function actually puts the media data into the media buffer:

```
  function doAppendStreamingData(data) {
      var uint8array = new window.Uint8Array(data);
      receiverBuffer.appendBuffer(uint8array);
  };
```

15. Implement a function that will stop playing back the media when the media data is over:

```
function doEndStreamingData() {
    recvMediaSource.endOfStream();
};
```

16. Create a function to send media data chunks to the remote peer. We will use the JSON format for such messages to declare type and data fields:

```
function pushChunk(data) {
    var msg = JSON.stringify({"type" : "chunk",
    "data" : Array.apply(null, data)});
    sendDataMessage(msg);
};
```

17. Implement a function that takes the received chunks and processes them:

```
function onChunk(data) {
```

18. We will put the first chunk into a cache and call the `doReceiveStreaming` function to prepare media components:

```
chunks++;
if (chunks == 1) {
    console.log("first frame");
    queue.push(data);
    doReceiveStreaming();
    return;
}
if (data.end) {
    console.log("last frame");
    doEndStreamingData();
    return;
}
```

19. In case the cache (queue) is not empty already, we will put the newly received chunk in the queue. That's because a non-empty queue means that we're receiving new chunks faster than we can process and show them:

```
if (receiverBuffer.updating || queue.length > 0)
queue.push(data);
```

20. In case the queue is empty, we can call the `doAppendStreamingData` function that will put the chunk in the media buffer, and the media data will be shown on the page:

```
else doAppendStreamingData(data);
};
```

So, you have the index page and the JavaScript library now. Put them both in the web server folder, and start the signaling server. Navigate your web browser to where the demo is accessible. Then navigate another web browser (or web browser tab) to the link at the top of the page; after this, peers will establish a direct connection.

At the bottom part of the page, you should see something similar to the following:

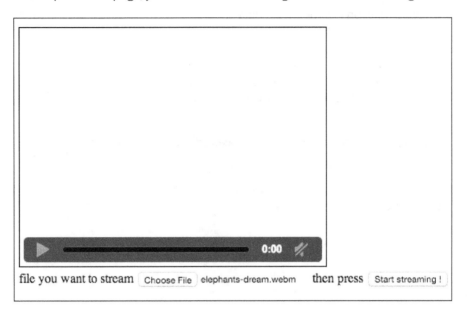

Note the buttons **Choose File** and **Start streaming!**. Click on the **Choose File** button and select the preloaded WebM media file. Then, click on the **Start streaming!** button. The web browser where you clicked the buttons will start reading the media file and streaming it to the second browser. So, on another browser window, you should see your media file playing.

In the following screenshot, you can see two browser windows: Chrome at the top and Firefox at the back. Here, I'm streaming the media file from Chrome to Firefox.

Note that this feature is in the beta stage, and you might need to make appropriate changes to make the demo work on other browser versions.

Another important note is that Firefox has disabled the `mediasource` component by default, so you should check that and enable it before using this recipe with Firefox. To do that, you should navigate to `about:config`, look for the **media.mediasource.enabled** option and set it to **true**. You can see this solution in the following screenshot:

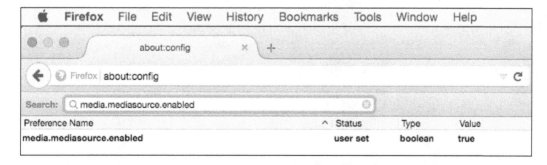

Note that Firefox starts playing immediately after it gets the first bytes of the media data. Chrome will wait until it gets all the media data and only then will start playing them. This behavior might be changed in other browser versions.

How it works...

The logic of this feature is simple. First of all, peers establish a direct connection and create data channel. Then, the sender (streaming peer) acts as shown in the following steps:

1. Reads the whole media file in the memory and creates a BLOB object.
2. Reads the BLOB object chunk by chunk, slices them into smaller blocks.
3. Sends the BLOB object chunk by chunk to the remote peer.
4. Repeats step 3 until the end of the media file.

On the other hand, another peer performs the following steps:

1. Creates a media source object. Prepares media buffer. Ties the objects with the video HTML object on the page.
2. Gets chunks from the remote peer and puts binary data in the media buffer, which is tied to the video object.
3. In case the streamer sends data faster than the receiver can process it, the receiver uses a queue to temporarily store the received media data.
4. Repeats steps 2 and 3 until there is some media data received from the remote peer.

Thus, the receiver plays back the video that is streamed by the remote peer.

See also

► This recipe actively uses WebRTC data channels. In the code, we considered only streaming-related important parts of code. For codes specific to data channels, refer to *Chapter 1, Peer Connections*, where this topic is explained in a more detailed way.

Index

Thank you for buying
WebRTC Cookbook

About Packt Publishing

Packt, pronounced 'packed', published its first book, *Mastering phpMyAdmin for Effective MySQL Management*, in April 2004, and subsequently continued to specialize in publishing highly focused books on specific technologies and solutions.

Our books and publications share the experiences of your fellow IT professionals in adapting and customizing today's systems, applications, and frameworks. Our solution-based books give you the knowledge and power to customize the software and technologies you're using to get the job done. Packt books are more specific and less general than the IT books you have seen in the past. Our unique business model allows us to bring you more focused information, giving you more of what you need to know, and less of what you don't.

Packt is a modern yet unique publishing company that focuses on producing quality, cutting-edge books for communities of developers, administrators, and newbies alike. For more information, please visit our website at www.packtpub.com.

About Packt Open Source

In 2010, Packt launched two new brands, Packt Open Source and Packt Enterprise, in order to continue its focus on specialization. This book is part of the Packt open source brand, home to books published on software built around open source licenses, and offering information to anybody from advanced developers to budding web designers. The Open Source brand also runs Packt's open source Royalty Scheme, by which Packt gives a royalty to each open source project about whose software a book is sold.

Writing for Packt

We welcome all inquiries from people who are interested in authoring. Book proposals should be sent to author@packtpub.com. If your book idea is still at an early stage and you would like to discuss it first before writing a formal book proposal, then please contact us; one of our commissioning editors will get in touch with you.

We're not just looking for published authors; if you have strong technical skills but no writing experience, our experienced editors can help you develop a writing career, or simply get some additional reward for your expertise.

WebRTC Blueprints

ISBN: 978-1-78398-310-0   Paperback: 176 pages

Develop your very own media applications and services using WebRTC

1. Create interactive web applications using WebRTC.

2. Get introduced to advanced technologies such as WebSocket and Erlang.

3. Develop your own secure web applications and services with practical projects.

WebRTC Integrator's Guide

ISBN: 978-1-78398-126-7   Paperback: 382 pages

Successfully build your very own scalable WebRTC infrastructure quickly and efficiently

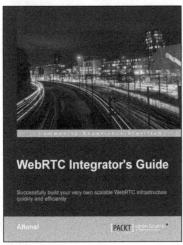

1. Build a feature-rich WebRTC client application and set up an Intelligent Network.

2. Use simple JavaScript APIs to enable web browsers with real-time communication (RTC) capabilities.

3. Make a real-time communication architecture through various modules, illustrations, and explanations with example code snippets.

Please check **www.PacktPub.com** for information on our titles

www.ingramcontent.com/pod-product-compliance
Lightning Source LLC
Chambersburg PA
CBHW082118070326
40690CB00049B/3669